Springer Series in Materials Science

Volume 230

The Springer Series in Materials Science covers the complete spectrum of materials physics, including fundamental principles, physical properties, materials theory and design. Recognizing the increasing importance of materials science in future device technologies, the book titles in this series reflect the state-of-the-art in understanding and controlling the structure and properties of all important classes of materials.

More information about this series at http://www.springer.com/series/856

Rostislav A. Andrievski · Arsen V. Khatchoyan

Nanomaterials in Extreme Environments

Fundamentals and Applications

 Springer

Rostislav A. Andrievski
Institute of Problems of Chemical Physics
Russian Academy of Sciences
Chernogolovka
Russia

Arsen V. Khatchoyan
Institute of Structural Macrokinetics
and Materials Science
Russian Academy of Sciences
Chernogolovka
Russia

ISSN 0933-033X   ISSN 2196-2812   (electronic)
Springer Series in Materials Science
ISBN 978-3-319-25329-9   ISBN 978-3-319-25331-2   (eBook)
DOI 10.1007/978-3-319-25331-2

Library of Congress Control Number: 2015955386

Springer Cham Heidelberg New York Dordrecht London

Printed on acid-free paper

Springer International Publishing AG Switzerland is part of Springer Science+Business Media
(www.springer.com)

Foreword

Most of today's technologies depend on the availability of materials with specific properties, such as high strength, light weight, specific electronic properties, superconductivity, and many others. The development of all of these materials may be divided into two periods of time. The first period is characterized by the discovery of a new effect or the proposal of a conceptually new idea. This initial period involves a small number of scientists who focus primarily on exploring the discovery or the new concepts. Their work is in most cases characterized by close interactions among the scientists involved.

During the second period of development, the number and the breadth of the studies performed and the diversity of methods applied increases significantly. This change results in a rapidly growing body of knowledge about the new field. This knowledge provides the basis for the development of a fundamental understanding of the results obtained so far. On the other hand, this fundamental understanding is the prerequisite for the following two developments: (1) the design of further studies deepening and broadening our understanding and (2) the recognition of conceivable technological applications.

In the history of nanomaterials, the two periods of development mentioned above are clearly born out. In 1979, the new concept of nanomaterials was proposed. The basic idea was to create crystalline materials consisting of a large (50 % or more) volume fraction of intercrystalline interfaces. As the arrangements of atoms in intercrystalline interfaces differ from the ones in the corresponding crystals and glasses, it was suggested that nanostructured materials would open the way to materials with new atomic structures and hence new properties. When this concept was confirmed by the pioneering studies during the early 1980s, the number of publications increased rapidly. In fact, today more than 300 papers related to all kinds of nanoscale effects published daily and in 2014 more than 120,000 publications in this young field of Science/Technology were retrieved by the Science Citation Index Expanded.

As a consequence, further progress in the area of nanomaterials depends critically on the availability of overviews in which the wide spectrum of the rapidly growing number of new results is presented within a limited number of pages, and in a critical way. In fact, the crucial tasks of these overviews have to be (1) to focus on the new insights gained by the newly published papers and (2) to create the "shoulders" on which the next generation of researchers can safely stand, when they try to look beyond the presently existing limits. Clearly, overviews of this type have to cover a wide spectrum of aspects, extending from different atomic, electronic, and chemical structures of nanomaterials all the way to thermodynamic aspects (such as questions of their stability, their kinetic properties, etc.), the various types of interactions with their environments (e.g., optical, chemical, electric, and other ones) and, moreover, it has to include the wide variety of specific nanostructured solids, such as fullerenes, nanotubes, polymeric nanostructures, nanoglasses, etc. So far, no overview seems available covering all of these aspects. The overview presented by Prof. Rostislav Andrievski and Dr. Arsen Khatchoyan is a first and important step in this direction. Their ability to review a significant portion of the field of nanomaterials is clearly evidenced by the contributions they have made in the past. In fact, Prof. R. Andrievski is well known as an editor and an author of international reputation in the area of nanomaterials and Dr. A. Khatchoyan made himself a name as a translator. In view of the rapid growth of this field, a review of this kind seems of particular importance for the next generation of scientists who try to work in that area. For all of them, there is still "plenty of room at the bottom." However, it will be the task of the further generations to find their way through the "room at the bottom" and to discover some of the fascinating effects that are still waiting there for all of us. Certainly, important navigational tools for these expeditions into "the room at the bottom" will be overviews of the kind presented by Prof. Andrievski and Dr. Khatchoyan.

Karlsruhe, Germany   Prof. Dr. Dr. h.c. mult Herbert Gleiter

Contents

About the Authors

Prof. Rostislav A. Andrievski Dr. Sci. (Materials Science) is a Principal Scientist at the Institute of Problems of Chemical Physics, Russian Academy of Science. He was born in 1933. He graduated from Kiev Polytechnic Institute, Department of Metal Science and Powder Metallurgy, and received his Ph.D. here in 1959 (Thesis "Sintering of Metal Powders"). From 1959 to 1962, Dr. Andrievski worked as a Senior Scientist at the Institute of Metal-Ceramics and Special Alloys, Ukrainian Academy of Sciences (Kiev). From 1963 up to 1976, Dr. Andrievski worked as a Head of Department of Materials Science at the Podolsk Technological Institute. He received his Doctorate at the Moscow All-Union Institute of Inorganic Materials in 1969 (topic of dissertation "Diffusion and Creep in Carbides and Nitrides"). From 1977 to 1984, he worked as Professor at the Moscow Lomonosov Institute of Fine Chemical Technology, Department of Refractory Compounds. In 1984, he was elected Corresponding Member of the Kirghiz Academy of Sciences and worked as Deputy Director of the Institute of Physics in Frunze (now Bishkek). Since 1990 and currently, Dr. Andrievski works at the Institute of Problems of Chemical Physics located in Chernogolovka, near Moscow. His main scientific interests are physics, chemistry, and technology of materials-based high-melting point compounds (such as carbides, nitrides, borides, oxides, etc.) as well as hydrides, especially in nanocrystalline state.

He has authored over 470 publications including 11 monographs (in Russian), 70 reviews, and 90 papers in reviewed foreign Journals and Proceedings. According to SCOPUS data, his Hirsch Index is 20 and Citation Index CI is 1450. He is a Supervisor of 33 Ph.D. theses and a Consultant of six Doctors of Sciences. Dr. Andrievski is a member of four International Editorial Boards and four of those are Russian, as well as a Guest Editor of three special Issues on nanomaterials in International Journals. He is a Member of the Materials Research Society (USA) and a Full Member of the International Institute for the Science of Sintering

(Serbia). Professor Andrievski was invited as a guest lecturer in a number of foreign Universities, such as Limoges University (France), Coimbra University (Portugal), Ben-Gurion University (Israel), Harbin Institute of Technology (China), and Indian Institute of Technology Kanpur (India). He is a well-known specialist in the field of size effects in nanomaterials and their stability at extremes. He presented his scientific results at all 12 International NANO Conferences—from Cancun-first (1992) to Moscow-twelve (2014).

Arsen V. Khatchoyan Ph.D. (Physics) is a Senior Scientist at the Institute of Structural Macrokinetics and Material Sciences, Russian Academy of Sciences. He was born in 1942 and graduated from the Erevan State University (Department of Nuclear Physics) in 1965. In 1966–1969, he was a postgraduate student in the Moscow Laboratory of the well-known Prof. L.S. Polak. In 1971, he received his Ph.D. at the Moscow Physical Technical Institute (Ph.D. thesis: "Rate Constants of Non-Equilibrium Plasmochemical Reactions"). In 1970–1976, he worked as a Scientist/Senior Scientist at the Institute of Physics and Mathematical Center of the Armenian Academy of Science; in 1974–1979, he worked as a lecturer of Physical Faculty at the Armenian Pedagogy Institute (courses on thermodynamics and statistical physics). From 1980, he is a Senior Scientist at the Chernogolovka Division of the Moscow Chemical Physics Institute (after reorganization and now the Institute of Structural Macrokinetics and Material Science, Russian Academy of Sciences). The scopes of his scientific interest are plasmochemistry, Boltzmann equation, and nonequilibrium statistical physics (about 40 scientific publications).

He is well known also as a translator/editor of the technical literature in general, and, from 1980, worked at MIR and other Russian Publishing Houses. He is the author of the first Japanese–Russian Chemical Dictionary (Moscow, Russkii Yazyk Publ. 1986, 35,000 terms) and the scientific editor of the first Japanese–Russian Radio-Electronics Dictionary (Moscow, Russkii Yazyk Publ., 1981). He is the translator and editor (partly together with Prof. R.A. Andrievski) of more than 30 translated books concerning modern physics, chemistry, and nanotechnology (some books are listed below):

1. V. Stiller. Arrhenius Equation and Non-Equilibrium Kinetics. Moscow, MIR Publ., 2000, p. 179
2. Nanotechnology Research Directions: IWGN Workshop Report. Vision for Nanotechnology R&D in the Next Decade (Eds.: M.C. Roco, R.S. Williams, and P. Alivisatos). Moscow, MIR Publ., 2002, p. 292
3. J. Altmann. Military Nanotechnology. Moscow, Technosphera Publ., 2006, p. 424

4. N. Kobayashi. Introduction in Nanotechnology. Moscow, BINOM Publ., 2007, p. 134 (translation from Japanese)
5. Lynn E. Foster. Nanotechnology. Moscow, Technosphera Publ., 2008, p. 340
6. E. Roduner. Nanoscopic Materials. Size-Dependent Phenomena. Moscow, Technosphera Publ., 2010, p. 352
7. Nanostructured Coatings (Eds.: A. Cavaleiro and J.T.M. De Hosson). Moscow, Technosphera Publ., 2011, p. 752

Acronyms

APT	Atom probe tomography
ARB	Accumulative roll bonding
BCC	Body-centered cubic
BET	Brunauer, Emmett, and Taylor
CG	Coarse-grained
CI	Citation index
CR	Cold rolling
CVD	Chemical vapor deposition
CVI	Chemical vapor infiltration
DFT	Density function theory
dpa	Displacements per atom
DTA	Differential thermal analysis
EBSD	Electron back-scattering diffraction
ECAP	Equal-channel angular pressing
EDS	Energy dispersion spectroscopy
FCC	Face-centered cubic
FMRR	Fast multiple rotation rolling
GBs	Grain boundaries
GG	Grain growth
GIAXRD	Glancing-incident angle X-ray diffraction
GS	Grain size
HE/R	Hot extrusion and rolling
HIP	Hot isostatic pressing
HMPC	High-melting point compounds
HPT	High-pressure torsion
HRSEM	High-resolution scanning electron microscopy
HRTEM	High-resolution transmission electron microscopy
IAV	Interstitial atoms and vacancies
ITER	International Thermonuclear Experimental Reactor
MA	Mechanical alloying
MD	Molecular dynamics
NLS	Nanolaminated structure

NMs	Nanomaterials
ODS	Oxide dispersion strengthened
PKA	Primary knocked-out atoms
SAED	Selected area electron diffraction
SEM	Scanning electron microscopy
SFT	Stacking-fault tetrahedra
SMAT	Surface mechanical attrition treatment
SMGT	Surface mechanical grinding treatment
SPD	Severe plastic deformation
SRT	Surface rolling treatment
TBs	Twinned boundaries
TEM	Transmission electron microscopy
TGA	Thermo gravimetric analysis
TJs	Triple junctions
UFG	Ultrafine-grained
XPS	X-ray photoelectron spectroscopy
XRD	X-ray diffraction

Chapter 1
Introduction

Abstract In this introductory chapter, attention is drawn to the rapid growth of information flow in the field of nanomaterials (NMs) and nanotechnologies. This growth has since the mid 90-ies almost exponential in nature, far ahead of the information accumulation in other areas of materials science and technology. Briefly, the NMs concept put forward in the works of Prof. H. Gleiter and his followers is described. There has recently been a heightened interest to the problem of the behavior of substances and materials in extreme conditions. The definition of extreme conditions with regard to NMs due to their peculiarity as unstable objects is specified. The main topics of monograph, such as the NMs behavior in extreme conditions of high temperatures, irradiation with ions and neutrons, as well as high mechanical and corrosion effects, are shortly described.

Several years ago, the MRS Bulletin Editor David Eaglesham published paper-question "The nano age?" and illustrated his mind by curves describing the publication activity in the different fields of the material science investigations. As shown in Fig. 1.1a (adapted from [1]), the advanced development of the NMs investigations is evident. Prolonging the curves of such nano-paper growth up to 2014 (according to estimations of Science Citation Index Expanded), we can see practically exponential growth of nano-information up to the present (Fig. 1.1b).

The NMs information growth (in that the nanotechnology one is also included) can be compared with tsunami! In general, the growth can be explained by both the widening of investigations/applications and the expansion of studied objects over, such as nanotubes, graphene, quantum dots, nanoglasses, etc. In addition, an interdisciplinary nature of nanoscience also promotes information widening. An essential role in the studies increase should be attributed to natural tendency of getting new results concerning the NMs behavior in different conditions. A thorough analysis of materials and substances evolution under extreme actions and conditions has recently become significant (e.g., [2–11]), at least because of two important circumstances. Firstly, the operating conditions of many modern devices and units are progressively changing towards increasing thermal, mechanical, radiation, corrosive, and other combined impacts of the operational environment, not to mention the time of service durations. Such a tendency requires

© Springer International Publishing Switzerland 2016
R.A. Andrievski and A.V. Khatchoyan, *Nanomaterials in Extreme Environments*,
Springer Series in Materials Science 230, DOI 10.1007/978-3-319-25331-2_1

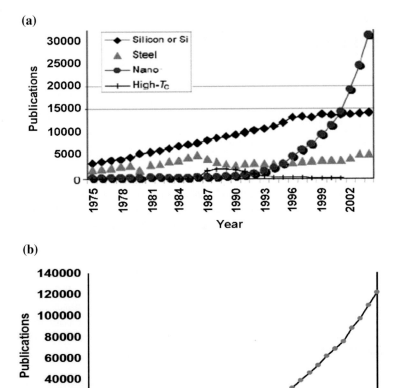

Fig. 1.1 Publications number over time for **a** different materials in the 1975–2005 period and **b** NMs in the 1980–2014 period

a special attention to the behavior and/or stability of NMs under listed extremes, since the search for new NMs must consider that they could withstand extreme environments. Secondly, a progress in astrophysics, cosmology, and geophysics leads to the new fundamental concepts and investigations connected with the substance extreme states as well as high energy and high mass densities creation [2–5]. In spite of the materials and substances tests at extremes are carried out at different pressure, temperature, and other parameters, test-modes sometimes overlap and undoubtedly complement each other.

Broadly speaking, the term "extreme conditions" in relation to NMs has features in comparison with conventional substances. It is well known that practically all NMs are far from the equilibrium, because many factors, such as the presence of numerous grain boundaries (GBs) as interfaces, triple junctions (TJs), residual stresses, segregations, and non-equilibrium phases, provide additional contributions to the Gibbs free energy (ΔG). In a greater or lesser extent, this commitment must always lead to the equilibrium violation, albeit simultaneously it creates the grounds for improvement of the NMs physical, mechanical, and operating characteristics. As will be shown below, the NMs (namely by virtue of their structure features) under extremes can be either more or less stable as compared with conventional coarse-grained (CG) counterparts. The situation is non-trivial due to numerous varying and discrepant results available in literature that demand analysis and generalization. The presentation given below assumes that the readers are acquainted with the basic data related to the NMs themselves and their manufacturing methods that were described in detail elsewhere [12–15].

It is useful to remind that historically the nano age in material science is associated with the NMs concept suggested by Gleiter and his followers [16–19]. Widely cited (citation index CI more than 2,900) review of Gleiter "Nanostructured materials" [20], published in 1989, played a major role in the expansion of research and publications (see Fig. 1.1) in this area. At present, Gleiter's concept lays the foundation of many modern theories in material science in general. The main issue of the approach concerns a principal role of numerous interfaces, which always are broadly presented in NMs. Theoretically, it was also proposed that solid body properties can be significantly changed by both modification of their crystal/electronic structure and new methods of doping by various elements regardless of their chemical nature or atomic size.

Apparently, the first estimation of the volume fractions of total interfaces, GBs, and TJs in NMs was published by Aust et al. [21]. The calculations were based in assumption that grains have the shape of 14-sided tetrakaidecahedras while the related intergrain spacing widths are about 1 nm. As can be seen from results [21] (see also later Fig. 5.1), in conventional CG crystalline materials the fraction of all interfaces (especially of TJs) is vanishingly small and can recorded when grain size (GS) is below of ~ 100 nm. The TJs predominant volume content can be revealed at the nanograin sizes below of 5 nm only.

The consolidated (non-polymeric) NMs classification has been also suggested by Gleiter [22]. This classification involves of 12 types of nanostructures, i.e., 3 categories of shapes (layers, rods and equiaxed grains) and 4 families depending on the chemical composition of GBs and inclusions (Fig. 1.2) [22]. Therefore, there are many types of interfaces in NMs. Theoretical ideas introduced in publications [16–20, 22] were later extended and developed that allowed formulating new principles and methods. It must be noted that now there are principally new interesting ideas of processing nanocomposites from oppositely charged nanocrystallites [23] and nanoglasses with tunable structure [24–27]. These approaches open a wide prospect of creating different new NMs thus allowing "to synthesize materials with properties beyond the today limitations" [24].

However, a comprehensive description of the numerous nanostructures behavior and stability under highly variable conditions and actions remains practically impossible, first and foremost due to the lack of important data. In the current

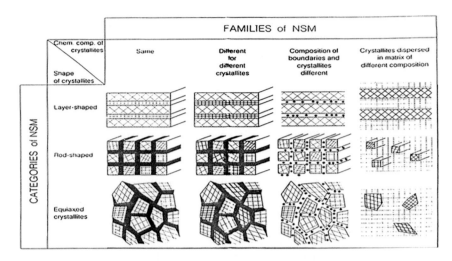

Fig. 1.2 Gleiter's classification of nanostructured materials (NSM): 3 categories of shapes (layers, rods and equiaxed grains) and 4 families depending on the chemical composition of GBs and inclusions (adapted from [22])

monograph, the main attention will be focused on the consolidated NMs-based metals, alloys and high-melting point compounds (HMPC), such as carbides, borides. nitrides, oxides, etc., for which the extremes influence has been extensively studied and even then only for two or three types of nanostructures (layer shaped objects, equiaxed crystallites and those with dispersed inclusions; see Fig. 1.2). The extreme conditions will involve thermal, irradiation, deformation, and corrosion impacts applied in the most realistic conditions, once characteristic for the present and future exploitation of NMs. From here on, we shall discuss not only NMs with the upper GS conditional limit of ~ 100 nm, but in many cases adjoining (related) materials with ultra-fine grained (UFG) structure (the GS value from about 100 to ~ 1000 nm) will be included for comparison as well.

References

1. Eaglesham DJ (2005) The nano age? MRS Bull 30:260
2. Fortov VE (2010) Ekstremal'nye Sostoyaniya Veshchestva (Extreme States of Matter). FIZMATLIT, Moscow (in Russian)
3. Hemley RJ, Crabtree GW, Buchanan MV (2009) Materials in extreme environments. Phys Today 62(11):32–37
4. Misra A, Thilly L (2010) Structural metals at extremes. MRS Bull 35:965–972
5. Boldyreva EV (2012) Supramolecular systems in extreme environments. Herald Russ Acad Sci 82:982–991
6. Fortov VE, Mintsev VB (2013) Extreme states of matter on the earth and in cosmos: is there any chemistry beyond the megabar? Russ Chem Rev 82:597–615

7. Bourne N (2013) Materials in mechanical extremes – fundamentals and applications. Cambridge University Press, New York
8. Low IM, Sakka Y, Hu CF (eds) (2013) MAX phases and ultra-high temperature ceramics for extreme environments. IGI Global, Hershey
9. Bini R, Schetto V (2014) Materials under extreme conditions. molecular crystals at high pressure. World Scientific Publishing, New Jersey
10. Fahrenholtz WG, Wuchina EJ, Lee WE et al (eds) (2014) Ultra-high temperature ceramics. Materials for extreme environment applications. The American Ceramic Society, Wiley, New Jersey
11. Andrievski RA (2014) Nanostructures under extremes. Phys-Usp 57:945–958
12. Poole ChP, Owens FJ (2003) Introduction to nanotechnology. Wiley, Weinheim
13. Koch CC, Ovid'ko IA, Seal S et al (2007) Structural nanocrystalline materials: fundamentals and applications. Cambridge University Press, Cambridge
14. Cavaleiro A, De Hosson JT (eds) (2006) Nanostructured coatings. Springer, Heidelberg
15. Valiev RZ, Zhilyaev AP, Langdon TG (2014) Bulk nanostructured materials: fundamentals and applications. Wiley, Weinheim
16. Marquardt P, Gleiter H (1981) Herstellung und eigenschaften von mikrokristallinen festkörpern. In: Heinicke W (ed) Proceedings of the Deutsche Physikalische Gesellschaft, Verhandlungen DPG (VI), vol 16. Physik Verlag GmbH, Weinheim, p 375
17. Gleiter H (1981) Materials with ultrafine grain size. In: Hansen N, Leffers T, Lilholt H (eds) Deformation of polycrystals. RISO Nat Lab, Roskilde, pp 15–21
18. Birringer R, Gleiter H, Klein H-P et al (1984) Nanocrystalline materials: an approach to a novel solid structure with gas-like disorder? Phys Lett 102:365–369
19. Birringer R, Herr U, Gleiter H (1986) Nanocrystalline materials—a first report. Trans Jap Inst Met Suppl 27:43–52
20. Gleiter H (1989) Nanostructured materials. Progr Mater Sci 33:223–315
21. Palumbo G, Erb U, Aust K (1990) Triple line disclination effect on the mechanical behavior of materials. Scr Met Mater 24:1347–1350
22. Gleiter H (1995) Nanostructured materials: state of the art and perspectives. Nanostr Mater 6:3–14
23. Gleiter H, Weissmüller J, Wollersheim O et al (2001) Nanocrystalline materials: a way to solids with tunable electron structure and properties? Acta Mater 48:737–745
24. Gleiter H (2008) Our thoughts are ours, their ends none of our own: are there ways to synthesize materials beyond the limitation today? Acta Mater 56:5875–5893
25. Gleiter H (2013) Nanoglasses: a new class of nanocrystalline materials. Beilst J Nanotech 4:517–533
26. Gleiter H, Schimmel Th, Han H (2014) Nanostructured solids – from nano-glasses to quantum transistors. Nano Today 9:17–66
27. Andrievski RA (2013) Metallic nano/microglasses: new approaches in nanostructured materials science. Phys-Usp 56:261–268

Chapter 2
Grain Growth and Nanomaterials Behavior at High Temperatures

Abstract Current developments in kinetic and thermodynamic stabilization of grains in NMs-based metals, alloys and HMPC at high temperatures are generalized and discussed in detail. Special attention is paid to a possible quantitative estimation with using the regular solution approximation by considering both inner regions of nanograins and their interfaces. Recent data on abnormal grain growth are also considered. Practical application examples concerning bulk and film/coating objects are given and some unsolved problems are presented.

2.1 General Considerations

Many of the below described structural and functional NMs are specially destined to be used at high temperatures (for example, heat-resistant nanocomposites and tools, materials for friction assemblies and electric contacts, catalysts, emitters, sensors, etc.). For these objects, a behavior under heating accompanied by a grain growth (GG) and other processes is obviously very important. The main stimulus for grain GG is the ΔG decrease at the sacrifice of GBs component. It is known that the neighbor grains number in honeycomb-type structures with the TJs angles 120° (so-called equilibrated TJs) is equal to six and it is presumed that such structures can remain thermally stable for a very long time. Thus, the basic drive forces of GG processes are the GBs curvatures and neighbors number deviations in both sides though a growth is also connected with the GBs mobility. The theoretical description of grain growth in the usual CG materials and NMs is given by Ovid'ko in Chap. 3 of monograph [1].

In the majority of cases, CG objects are characterized by a cooperative homogeneous grain displacement and merging when their lognormal or normal GS distributions preserve (Fig. 2.1b). In NMs grain growth proceeds owing to their rotation as well (Fig. 2.1c). Also, an abnormal grain growth is observed which is accompanied with the appearance of CG seats or centers absorbing their smaller neighbors as it is in Fig. 2.2 with possible formation of a new bimodal GS distribution. Such abnormal grain growth is also characteristic in some NMs produced by

© Springer International Publishing Switzerland 2016
R.A. Andrievski and A.V. Khatchoyan, *Nanomaterials in Extreme Environments*,
Springer Series in Materials Science 230, DOI 10.1007/978-3-319-25331-2_2

(b)   **(a)**   **(c)**

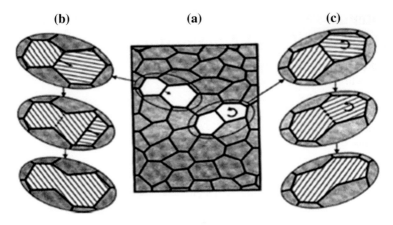

Fig. 2.1 Scheme of grain growth: **a** Initial state. **b** Migration and boundary merging. **c** Rotation and boundary merging (adapted from [1])

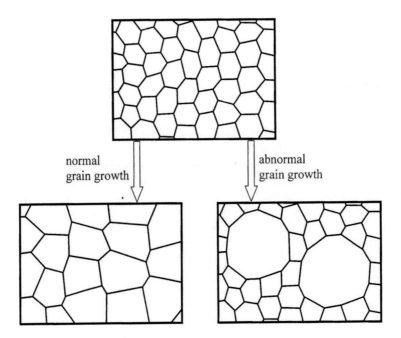

Fig. 2.2 Scheme of normal grain growth and abnormal that (adapted from [1])

severe plastic deformation (SPD) methods, where heating (besides grain growth) is attended with the removal of the crystal lattice micro- and macro distortions, as well as other relaxation effects (see [2, 3] for more details).

The mechanism of the rotational grain growth in NMs is imperfectly studied and therefore some recent investigations in this direction should be marked. It was shown through using molecular dynamics (MD) simulation that in nanostructures GBs migration and grains rotation can occur simultaneously, especially for small GS [4]. The velocity of grain rotation and its contribution to the general GG became less important as a result of GBs migration, which caused an increase in GS during grain growth. It was also mentioned that the rotation velocity depends on the conjunction geometry of touching grains [5]. By the example of Y_2O_3, it was shown that during the rotation of 20–200 nm grains low-energy/low-angle boundaries are formed in the 600–1000 °C range [6]. An interesting example of a collective grain rotation was observed in TiN and NiO nanocrystalline layers irradiated by high energy Au ions ($E = 360$ MeV) [7].

2.2 Some Theoretical Approaches and Modeling

As applied specially to consolidated NMs, in order to retard the nanograin growth (and correspondingly to prevent the nanostructure degradation), one can use the following methods: retardation by the second phase inclusions or by pores; change of the initial GS; decrease in the grain mobility by alloying (including segregation formation on GBs) and the decomposition of high-temperature spinodal. The GB migration velocity (v) under the action of own curvature is usually described by the traditional relationship:

$$v = M \cdot P = M_o \exp\left[\frac{-Q_m}{RT}\right] \cdot \frac{2\gamma_g}{r_g}, \tag{2.1}$$

where M is the GBs mobility, M_o is a pre-exponential factor, P is the driving force, Q_m is the activation energy of the boundary moving process, γ_g is the GB energy, and r_g is the average grain radius. By convention, the grain retardation process is usually named a *kinetic* approach (when the process is a consequence of the driving force) or a *thermodynamic* one (when it is a result of the driving force decrease at the sacrifice of the GB energy reducing). Theoretically, the both approaches have been derived into a different degree, but the GG retardation due to the micro/nanoinclusions is more widely used in material science. The GG retardation due the inclusions (so-called Zener pinning mechanism) is described by the equation:

$$P_z = \frac{3V_r\gamma_g}{2r_i}, \tag{2.2}$$

where P_z is the Zener drag force and V_r is the volume fraction of randomly distributed spherical inclusions with radius r_i. It is clear from (2.2) that the effective GG retardation can be achieved by increasing the number of inclusions and/or decreasing their radius.

In the frame of kinetic approach, the influence of nanoinclusions, pores, TJs, and quaternary points on thermal stability was analyzed by Novikov, Shwindlerman et al.

(e.g., [8–12]). From their results, it is worthy to mention that the role of TJs and quaternary points in retardation process can be described by the expression:

$$v = \frac{\gamma_g MK}{1 + 1/\Lambda + 1/\Lambda^*},$$

(2.3)

where K is the GBs curvature, $\Lambda = aM_T/M$, and $\Lambda^* = a^2 M_Q/M$, where a is the spacing between triple lines, and M_T and M_Q are the mobilities of TJs and quaternary points, respectively [11]. The numerical estimations using expression (2.3) have demonstrated [10] that the junction contribution in the nanostructure stabilization increases with decreasing initial GS. It is of interest that the process is especially well detectable at low temperatures, but after heating some inversion is detected and the boundaries with junctions become more mobile.

The thermodynamic analysis allowed determination the sign and value of the TJs linear extension, which was found to be positive and equal to $(6.03 \pm 3.0)\ 10^{-8}$ J/m [12]. It its turn, this value led to a more correct estimation of the GG driving force, and it was shown for Cu that a critical mean GS (below which the TJs values must be properly accounted in the estimation of boundary migration) is about 55 nm.

The numerical models of nanoparticle movement during GBs migration have revealed some dualism in a nanostructure evolution, which consists in the fact that the nanoparticles presence (depending on their volume content and total GBs migration) not only can decrease grain mobility, but also assists its increase leading to an abnormal grain growth [8, 9]. Consideration of various versions for the interaction of pores and GBs has shown that processes of grain growth and pore annihilation under annealing can proceed with different kinetic parameters depending on the component size and mobility [13].

To calculate the GBs mobility, thermodynamic studies of two-component nanosystems were carried out that allowed estimation of ΔG value for alloys with taking into account not only a concentration factor but also the presence of nanograins [14–17]. The key moment in these studies was to apply the well-known regular solution approximation to both bulk and GB regions of nanocrystals. Finally, the equation for the ΔG value took the form [15, 16]:

$$\Delta G = (1 - f_b)\,\Delta G_g + f_b \Delta G_b + z\psi f_b\left(X_b - X_g\right)$$
$$\left[(2X_b - 1)\omega_b - (zt)^{-1}\left(\Omega^B \gamma_g^B - \Omega^A \gamma_g^A\right)\right],$$

(2.4)

where ΔG_g and ΔG_b are the Gibbs free energies for inner grain regions (g) and GBs those (b); f_b is the volume fracture of GB regions defined by relation:

$$f_b = 1 - \left(\frac{L - t}{L}\right)^3,$$

(2.5)

where L is GS; t is the thickness of GB region (usually equals 0.5–1 nm); z is the coordination number for bulk material A; ψ is the fraction of atoms with interatomic bonds in GBs regions; X_b and X_g are the solute concentrations in the GB and inner regions which must satisfy the condition of the average concentration $X = f_b X_b + (1 - f_b) X_g$; Ω^A and Ω^B are the component atomic volumes of alloy A–B components (addition); γ_g^A and γ_g^A are the partial energies of their GBs; and ω_b is the interaction parameter for the GB regions, used in the regular solution approximation defined as a function of the paired interaction energy (E) by expression:

$$\omega_b = E^{AB} - \frac{E^{AA} + E^{BB}}{2} \tag{2.6}$$

The thermodynamic properties of ΔG surfaces and their minima can be calculated using these equations and varying L values, which allows finding characteristic for the most thermally stable nanoalloys. The different situations of ΔG minimization applied to GG and phase decomposition (variation of content, change in temperature and interaction parameters, absence of ΔG minimum, different combinations of mixing and segregation enthalpies, presence of metastable nanoalloy, formation of two-phase nanocomposites, and so on) were analyzed thoroughly in [16].

Figure 2.3 shows the segregation enthalpy ΔH_S dependence on the mixing enthalpy ΔH_M for various W-based alloys at 1100 °C [15].

The segregation and mixing enthalpies were calculated with the following formulae:

$$\Delta H_S = z \left[\omega_g - 0.5\omega_b - \frac{\left(\Omega^B \gamma_g^B - \Omega^A \gamma_g^A \right)}{2zt} \right] \tag{2.7}$$

Fig. 2.3 Stability map of tungsten-based alloys at 1100 °C: *1* and *2* are the stability regions of nanocrystalline and CG W alloys, respectively (adapted from [15])

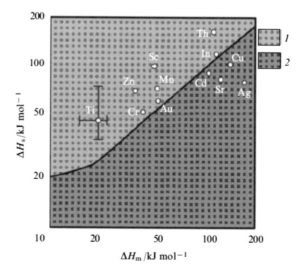

and

$$\Delta H_M = z\omega_g X(1 - X), \tag{2.8}$$

where ω_g is defined analogously to expression (2.6).

As indicated in Fig. 2.3, the nanocrystalline state is fixed in alloys for which the corresponding points lie above the $\Delta H_S = f(\Delta H_M)$ curve, whereas the CG objects correspond to the bottom part (2) of the diagram.

The same approach-based regular solution approximation for inner and GB regions [14] was used in [17], where the temperature effect was studied in Fe–Zr, Cu–Zr, Cu–Nb, and Ni–W alloys with a varying content of second component. The authors [17] have accounted the role of elastic contribution into the alloy formation in the ΔG calculations (apart from pure chemical interactions). The results showed that the most marked decrease in the grain growth under nanoalloy annealing is observed for zirconium added with small amount of iron.

Another method for GB and inner regions account was developed in [18–20] where authors took into consideration the difference in the excess free volume ΔV between GBs and bulk material. Finally, the relations for estimation of enthalpy H, entropy S and free energy G were derived as follows:

$$H = X_b H_b(\Delta V, T) + (1 - X_b)H_g(T), \tag{2.9}$$

$$S = X_b S_b(\Delta V, T) + (1 - X_b)S_g(T), \tag{2.10}$$

$$G = X_b G_b(\Delta V, T) + (1 - X_b)G_g(T), \tag{2.11}$$

where $\Delta V = [(V_b/V_g) - 1]$, V_b and V_g are volumes of primitive cells in the GB and inner regions, respectively. The formula for ΔV assumes some excess free volume in the systems due to non-ordered arrangement of atoms in GB regions ($V_b > V_g$). The ΔV values and thermodynamic functions for GB regions (including heat capacity and Debye temperature) are taken with regard to experimental data. The method has been used to estimate the thermal stability and phase transformations of some Sm–Co system compounds, such as $SmCo_5$, Sm_2Co_{17}, $SmCo_7$, etc. [18–22].

The thermodynamic stability of binary nanoalloys was analyzed also by a new Monte-Carlo modeling method [23], where it was proposed to correct the stwo-phase alloys stability map (see Fig. 2.3) by appending two additive regions describing the decomposition of CG alloys and duplex nanoalloys. Other theoretical studies estimated the spinodal decomposition conditions for the Al_2O_3–ZrO_2 and TiN–SiN quasi-binary systems as well as investigated the behavior of interfaces in multi-layer TiN/SiN/TiN nanocomposites [24–28]. In some studies [24–26], the thermodynamic approach was combined with the density function theory (DFT), which allowed calculation not only the thermodynamic characteristics (ΔG, ΔH, etc.), but also the electronic and crystal structure parameters, as well as the material elastic modulus. The calculations revealed that in the metastable TiN–SiN system the dome top of spinodal and bimodal decomposition occurs at about $T \sim 3200$ K, and below 1273 K the system practically is two-phased [24]. The peculiarities of interface thermal stability in various one-layer TiN/Si_3N_4 structures were investigating in the 0–1400 K range by using the so-called first principles method of MD [27, 28]. In particular, it was marked that the thermal instability of SiN cubic modification can be assigned to the vacancy formation in the silicon sublattice.

At the present time, it is difficult to give preference to any of described theoretical approaches notably, taking into account that the calculations by formulae (2.1–2.11), often are impossible because of the necessary thermodynamic data absence.

Furthermore, there is a general problem of the classical thermodynamics applicability to nanoscale objects. For example, Rusanov [29] supposes that the traditional notions, connected with surface energy, are acceptable as a whole only for isolated nanoparticles sizing over 10 nm. At sizes under 1 nm, any nanoparticle (or nanolayer) practically can acquire the surface properties, i.e., it transforms into a distinctive physical state (distinguished from a volume phase) and this fact requires a special approach. The size range of 1–10 nm is an intermediate one and in each particular case must be considered specially. Another lower limit for nanocrystal sizes was marked by Glezer [30], who pointed that the nanocrystallinity notion disappears with disappearing the symmetry elements characteristic for the given class of crystals. In other words, the meaning and significance of symmetry elements preserve only to the limit when their sizes are commensurable with the three coordination spheres. Then, the minimal critical crystallite size for the body-central cubic (BCC) and face-central cubic (FCC) structures is about 0.5 nm (α-Fe) and 0.6 nm (Ni), respectively. From the above mentioned data, it can be tentatively supposed that the GBs thickness (even with accounting the near boundary regions) is close to the lower limit of the classical thermodynamics applicability. In general, the question of the GB region behavior remains open and requires the further experimental and theoretical studies.

2.3 Main Experimental Results

2.3.1 Bulk Nanomaterials

The thermal stability of consolidated NMs was studied by many authors (see [1, 31–35] and references therein), therefore our attention below will be paid to recent results. Of them, new findings in support for the above described approaches seem to be the most important. In Table 2.1, the experimentally obtained GS values for the Cu–5 at.% Zr alloy samples annealed at 300–700 °C [36] are compared with values calculated by method described in [17]. It is easily seen that the difference is about 4–10 times, but these results can be accepted as satisfactory in regard to the model approximations [17].

Table 2.1 Experimental and calculated values of GS in annealed Cu–5 at.% Zr alloy

T (°C)	Grain size (nm)	
	Experimental values [36]	Calculated values [17]
300	9	35
500	15	55
700	22	230

The suggested presence of the nanocrystalline state in the W–Ti alloy in [15] (see Fig. 2.3) was verified in experiments with the W–20 at.% Ti powder (the initial nanograins size of 20 nm), which practically did not change after weekly annealing at 1100 °C. At the same annealing conditions, the GS in non-doped tungsten was about 600 nm [15, 37].

Thus, the developed theoretical methods [14–17] are acceptable for estimation of the metallic nanoalloy thermal stability. It is also important that the possible existence of W–20 at.% Ti alloys in a nanocrystalline state has been independently predicted by the Monte-Carlo calculations [37]. The study of titanium distribution in W–20 at.% Ti alloys nanostructure revealed its heterogeneous character because of the content variations in the range from 0 to 50 at.% [15, 37], i.e., under the considered conditions the alloy thermal stability can be explained not only by the free energy decrease, but also by the appearance of segregation inclusions and Zener pinning mechanism. The same situation (i.e., with the action of two mechanisms) can take place in the Cu–5 at.% Zr alloy annealed at 700 °C, when difference between the experimental and theoretical GS values (based on accounting one mechanism) was maximal (see Table 2.1).

To compare experimental findings of different studies concerning the effectiveness and congruence of both the kinetic and thermodynamic approaches for description of nanograin thermal growth retardation, the corresponding data were normalized to the melting temperature T_m in the homological temperature scale by Koch et al. [38]. From the other side, only the results for systems with a GS of under 100 nm in annealed samples were taken in account in [38]. Its authors only used most reliable data provided by TEM and XRD methods. The alloy-based Al, Mg, Cu, and Fe samples with nanoinclusions of Al_2O_3, AlN, ZrO_2, and Nb proved to be thermally stable at high homological temperatures up to $(0.75–0.85)$ T_m, whereas the alloy-based Ni, Co, Fe, Y, RuAl, and TiO_2 systems with additions of W, P, Zr, Fe, and Ca exhibited lower thermal stability up to $(0.35–0.65)$ T_m.

High thermal stability was fixed for copper alloys with tungsten and tantalum additives (10 at.%), prepared by a high-energy milling at cryogenic temperature of the initial powders with their following pressing and sintering [39, 40]. In other experiments, the initial GS of Cu–W alloys was of ~ 15 nm, but after annealing the grains had L of ~ 60 nm [37]. In Table 2.2, the results of more detailed investigation of the Cu–Ta alloys are presented. They show a high enough thermal stability of the objects.

For comparison, it is worthy to mark that the microhardness of Cu–10 at.% W alloys in the initial and annealed states equals about 3–2.6 GPa, correspondingly [38], i.e., it is lower than the values obtained in [40] (see Table 2.2). After detailed

Table 2.2 Evolution of GS and microhardness after annealing Cu–10 at.% Ta under pure H_2 gas (for 4 h) (adapted from [40])

Annealing temperature		Grain size (nm)		Microhardness (GPa)
(°C)	(T/T_m)	Cu phase	Ta phase	
20	0.22	6.7	6	4.8
540	0.6	21 ± 4	22 ± 5	∼3.6
770	0.77	54 ± 9	37 ± 9	n/d
900	0.87	111 ± 26	42.5 ± 25	3.4
1040	0.97	167 ± 50	99 ± 99	2.6

consideration of different mechanisms of thermal stability connected with TJs, doping, inclusions, segregations, etc., the authors [40] made a conclusion that the observed high thermal stability of such Cu–Ta alloys was connected with the formation and following decomposition of tantalum segregations at the GB regions, i.e., with some combined action of the thermodynamic and kinetic mechanisms. The latter comes at high temperatures as a result of the GB layer decomposition into inclusions (nanoclusters by terminology of authors [40]), which retards GG within a high temperature interval. This conclusion was verified by the MD calculations and sample structure experimental studies using TEM [41].

A significant role of the kinetic mechanism in the thermal stability due to the retarding action of the nanoinclusions was marked also for the Cu–Nb, Ni–Y, and Fe–Zr nanoalloys [42–44], but was not observed for Cu–Fe and Pd–Zr systems [43, 44]. The thermal stability of Cu-based nanoalloys is thoroughly considered in survey [35].

In addition to the above mentioned doping and inclusions, interfaces also retard grain growth, which can be adjusted during the NMs manufacturing processes. Figure 2.4 shows the temperature dependences of basic parameters, such as GS, lamellar spacing width, and hardness for copper samples prepared by both traditional methods of SPD and magnetron sputtering with the nanotwinned structure formation [45].

Here, one can see very weak changes in the lamellar thickness and high hardness for nanotwinned samples in distinction to the usual sufficient GS growth and hardness decrease with temperature for the common type nanostructures. This high thermal stability of the low-angle nanotwinned boundaries is related to their low GB energy, which is lower than that of ordinary high-angle boundaries by an order of magnitude. It is worthy to mention that the high electrical conductivity and wear resistance of such copper samples with a nanotwinned structure [46] is very important for various applications. Such advantages were described for nickel [47] and Cu–Nb nanocomposites [48] as well.

Let us notice that there are many situations when GG (in particular, an abnormal grain growth; see Fig. 2.2) in NMs cannot be prevented even at room temperature. The kinetics curves of the GG in nanocrystalline palladium at 20 °C are presented in Fig. 2.5, where coincidence of the experimental [49] and calculated [50] results seems satisfactory up to the beginning of abnormal grain growth (at a test duration

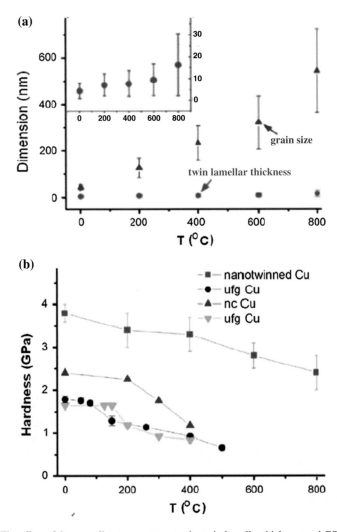

Fig. 2.4 The effect of the annealing temperature on the twin lamellar thickness and GS (**a**) as well as hardness (**b**) of sputtered nanotwinned Cu films and other, more traditionally prepared, ultrafine-grained (ufg)/nanocrystalline (nc) Cu specimens (adapted from [45]). The *inlet* shows the lamellar thickness change

of over 8 h). On the other hand, when the TJs retardation action is not taken into account, the discrepancy between theory and experiment becomes very striking. But neither the presence of many TJs nor the existence of residual pores (the initial GS was of ~ 5 nm and the porosity was of ~ 4 %) and admixtures (0.4 at.% H, 0.2 at.% N and 0.1 at.% O) can not prevent the abnormal grain growth, which leads to the transformation of the palladium nanostructure into ordinary microstructure with GS about ~ 10 μm after 2-month exposition at room temperature [49].

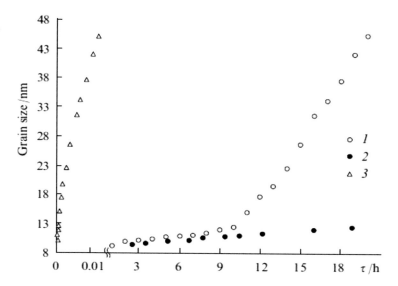

Fig. 2.5 Room temperature grain growth in nanocrystalline Pd: (*1*) Experimental results [49]; (*2* and *3*) calculated values with and without considering the TJs drag effect, respectively (adapted from [50])

An abnormal grain growth at high and room temperatures was observed in a number of experiments, including the TEM studies in situ [40, 53–55] for many other materials-based copper [40, 51], nickel [52–54], iron [55, 56], hard WC–Co alloys [57], and diamond [58]. It is worthy to notice that the listed materials were manufactured by various methods: the pulsing electrical deposition [54], pressing and sintering [49, 57], high-pressure torsion (HPT) at liquid nitrogen temperature [51], pulsing laser deposition [58, 59], intensive mechanical treatment (milling + mixing) or mechanical alloying (MA) [40, 55, 56], and sintering under high pressure ($T = 1400$ °C, $P = 6.8$ GPa) [58]. Naturally, the experiments with such wide diversity of objects and methods have revealed some new peculiarities in the initial and final states of material. In particular, the in situ TEM studies of nickel films made it possible to fix numerous defects in annealed samples, such as the presence of stacking-fault tetrahedra (SFT) and dislocation loops as well as the low-energetic TB formation [53, 54]. The authors [51] have revealed non-stability of copper nanostructure at room temperature and made a conclusion that the plastic deformation method at low temperatures is unpromising for NMs manufacture. The study in situ of the $Fe_{91}Ni_8Zr_1$ alloy thermal stability has shown that the abnormal grain growth begins above 700 °C and is related to the formation of γ-phase crystals with a FCC structure [56].

There are many new interesting findings concerning theoretical description of abnormal grain growth partially given above in Sect. 2.2 [8–10] and also in [59, 60], but the principal aspects of this effect are still to be solved (e.g., the role of defects and admixtures, prediction of change in the grain growth mechanism from normal to abnormal one, etc.). In conclusion of this subsection, it is worthy to mark that the recent studies of the nanoglass behavior under heating (for the Fe–Sc system samples) have revealed a relaxation process and some redistribution of excess free volume [61].

2.3.2 Nanostructured Films and Coatings

Above, some results have been presented demonstrating a high thermal stability of films with a nanotwinned structure and high-temperature nanocomposites (Fig. 2.4 [45] and [24–28]), but films and coatings themselves represent a very interesting and important field of investigations. Such nano-objects have been studied before (see [62–67] and references therein), but it is evident that they merit a more detailed consideration as convenient objects to get new important data. For example, the studies of one-/multi-layer nitride and nitride/carbide/boride films have already revealed a new class of superhard coating materials with hardness at the level of diamond and boron nitride (BN) (Table 2.3).

The results presented in Table 2.3 were obtained contemporaneously and independently of one another by investigators in many countries (USA, Sweden, former USSR, Austria, and Germany), and they all demonstrated the films high thermal stability as well as a role of the high-temperature spinodal decomposition in the nitride solid solutions. Figure 2.6 reveals that the thermal stability of new materials surpasses that of diamond, amorphization of which begins at ~ 800 °C. Also, it was found that the high thermal stability of (Ti, Zr)N films is connected with spinodal decomposition in the TiN–ZrN system with the nanocrystalline structure formation.

Table 2.3 Some first results for superhard films (adapted from [64])

Films	Microhardness (GPa)	Year	Authors[a]
One-layer films			
Ti(B, C)$_X$	~ 70	1990	Knotek et al.
Ti(B, N)$_X$	~ 60	1990	Mitterer et al.
B$_4$C	50–70	1992	Veprek
Multi-layer films			
TiN/VN	54	1987	Helmerson et al.
TiN/NbN	~ 78	1992	Andrievski et al.
TiN/NbN	48	1992	Shinn et al.
TiN/ZrN	~ 70	1992	Andrievski et al.

[a]See [64]

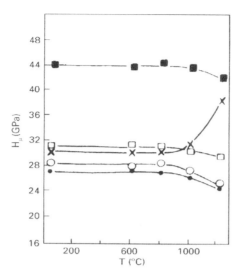

Fig. 2.6 The temperature effect on microhardness 2 μm thick nitride films. ● one-layer ZrN; ○ one-layer TiN; **x** solid solution (Ti, Zr)N disintegrated above 800 °C; □ TiN/ZrN (10 layers); ■ TiN/ZrN (20 layers) (adapted from [64])

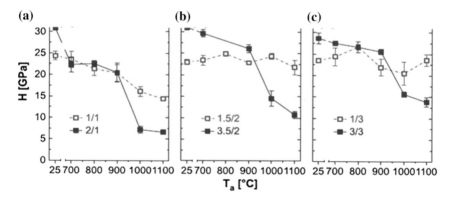

Fig. 2.7 The temperature effect on microhardness of CrN/AlN films with different thickness of individual layers: **a** 1 nm, **b** 2 nm, and **c** 3 nm thin AlN layers combined with CrN layers thicknesses of (**a**) 1 and 2 nm, (**b**) 1.5 and 3.5 nm, and (**c**) 1 and 3 nm (adapted from [68])

By now, new and more thermally stable films and coatings have been synthesized. For example, Fig. 2.7 demonstrates the temperature-microhardness dependences for CrN/AlN multi-layer films with varying layers thickness [68]. These results obviously indicate that the films with the AlN layers predomination (1.5/2 and 1/3) preserve their hardness after annealing at 1100 °C. The TEM and XRD studies allow suggesting that it is connected with a greater preservation of the

Fig. 2.8 The temperature
effect on hardness of the
TiAlN, TiAlSiN, TiAlCN,
and TiAlSiCN coatings
(adapted from [69])

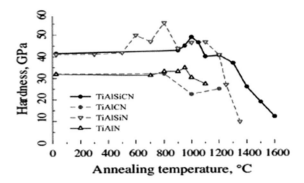

dissociating chromium nitride. In the experiments, the number of individual layers
was changing from 250 to 580 (with the same common film width about 1.5 μm).

The examination of the one-layer TiAlSiCN coatings after vacuum annealing in
the 900–1600 °C range showed that the high initial hardness ($\sim$42 GPa) changes
very little after annealing at 1300 °C with a "comb" like nanocomposite structure
preservation (Fig. 2.8) [69].

A high thermal stability after annealing in the 1000–1300 °C range was found
for the superhard nitride coatings-based high-entropy multicomponent Ti–Hf–Zr–
V–Nb alloys [70]. Now, it is evident that the evolution of high-temperature
structure and properties of these multicomponent nanostructured coatings are
defined by many various factors connected with relaxation, recrystallization, diffusion, and other processes, and we are only at the pioneering stage of such studies
(e.g., [69, 71, 72]).

2.4 Examples of Application

Today, the NMs high-temperature application can be realized predominantly as
films/coatings-based HMPC used for tools or friction units. As for the development
of tool NMs, Veprek's paper [73] devoted to search for new superhard materials
was of a great value. It was published more than 15 years ago and had a great
resonance (its citation index CI is over 860).

The proposed nc-(Al, Ti)N/a-Si$_3$N$_4$ composite coatings-based (Al, Ti)N
nanocrystallites in an amorphous matrix of silicon nitride exhibiting high hardness,
wear and oxidation resistances (Fig. 2.9) were realized in industry applications and
repeatedly improved [67, 74]. An optimal thickness of Si$_3$N$_4$ is now considered to
be about 0.3 nm (1 layer) [67].

In general, the main demands for tool materials can be formulated as follows: the
hardness and strength must be preserved at high temperatures, as well as the wear,
oxidation, and friction resistances. These characteristics were realized in the above
mentioned study of Shtansky et al. [69] by the example of TiSiCN coatings with Al

Fig. 2.9 Scheme of the *nc*-
(Al, Ti)N/*a*-Si$_3$N$_4$
nanostructure (**a**) and the
temperature evolution of its
hardness and crystallite size
(**b**) (adapted from [75, 76])

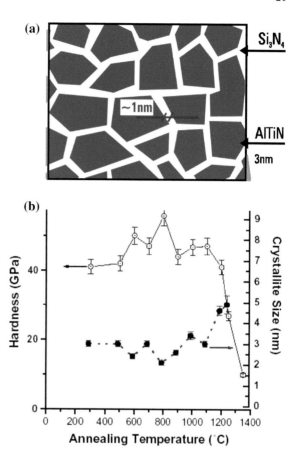

and Cr additions (see Fig. 2.8). A comparison of the thermal and oxidation stability
for various one-layer boron-nitride coatings (Ti–B–N, Ti–Cr–B–N, Ti–Si–B–N,
and Ti–Al–Si–B–N) was carried out in [77]. Some other combinations of extreme
actions including high temperatures, irradiation, deformation, and corrosion effects
will be described in the next Chaps. 3–5.

It is important to mark that various methods for manufacturing thermally stable
nanostructures [78, 79] are widely studied and carefully examined especially to
produce the materials-based steels, titanium, etc., generally used in machine con-
struction. As an example, to improve nanostructured stainless steel 316L, such
modern methods as HPT, equal channel angular pressing (ECAP), surface
mechanical attrition treatment (SMAT) and fast multiple rotation rolling (FMRR)
are used [78].

The intensive searches for rise in the NMs thermal stability are in progress, and
new practically important materials (such as alloys of Al, Mg, Ti, Cu, Ni, Fe, Nb,
Mo, and W as well as Ti and Sn oxides, silicon carbides, etc.) and results should be
forthcoming, particularly in manufacturing electric contacts, refractory items,

high-temperature sensors, etc. The GB engineering approach as applied to Ni–Fe-based superalloy 706 has been described by Detor et al. [80], but this study was realized only for CG objects and NMs approach still waits for its realization.

Different models of the thermodynamic GS stabilization in many metallic alloys reviewed and compared in an overview [81]. High thermal stability of superhard nanostructured lamellar (Ti, Zr)C was described in [82].

References

1. Koch CC, Ovid'ko IA, Seal S et al (2007) Structural nanocrystalline materials: fundamentals and applications. Cambridge University Press, Cambridge
2. Estrin Y, Vinogradov A (2013) Extreme grain refinement by severe plastic deformation: a wealth of challenging science. Acta Mater 61:782–817
3. Valiev RZ, Zhilyaev AP, Langdon TG (2014) Bulk nanostructured materials: fundamentals and applications. Wiley, Weinheim
4. Upmanyu M, Srolovitz DJ, Lobkovsky AE et al (2006) Simultaneous grain boundary migration and grain rotation. Acta Mater 54:1707–1715
5. Bernstein N (2008) The influence of geometry on grain boundary motion and rotation. Acta Mater 56:1106–1113
6. Chaim R (2012) Groan coalescence by grain rotation in nanoceramics. Scr Mater 66:269–271
7. Zizak I, Darowski N, Klaumünzer S et al (2009) Grain rotation in nanocrystalline layers under influence of swift heavy ions. Nucl Instr Meth Phys Res B 267:944–948
8. Novikov VY (2010) On grain growth in the presence of mobile particles. Acta Mater 58:3326–3331
9. Novikov VY (2012) Microstructure evolution during grain growth in materials with disperse particles. Mater Lett 68:413–415
10. Novikov VY (2008) Impact of grain boundary junctions on grain growth in polycrystals with different grain sizes. Mater Lett 62:2067–2069
11. Gottstein G, Shvindlerman LS (2005) A novel concept to determine the mobility of grain boundary quadruple junctions. Scr Mater 52:863–866
12. Zhao B, Gottstein G, Shvindlerman LS (2011) Triple junction effects in solids. Acta Mater 59:3510–3518
13. Klinger L, Rabkin E, Shvindlerman LS et al (2008) Grain growth in porous two-dimensional nanocrystalline materials. J Mater Sci 43:5068–5075
14. Trelewicz JR, Schuh CA (2009) Grain boundary segregation and thermodynamically stable binary nanocrystalline alloys. Phys Rev B 79:094112 (1–13)
15. Chookajorn T, Murdoch HA, Schuh CA (2012) Design of stable nanocrystalline alloys. Science 337:951–954
16. Murdoch HA, Schuh CA (2013) Stability of binary nanocrystalline alloys against grain growth and phase separation. Acta Mater 61:2121–2132
17. Saber M, Kotan H, Koch CC et al (2013) Thermodynamic stabilization of nanocrystalline binary alloys. J Appl Phys 113:063515 (1–10)
18. Xu WW, Song XY, Li ED et al (2009) Thermodynamic study on phase stability in nanocrystalline Sm–Co alloy system. J Appl Phys 105:104310 (1–6)
19. Song X, Lu N, Huang Ch et al (2010) Thermodynamic and experimental study on phase stability in nanocrystalline alloys. Acta Mater 58:396–407
20. Xu W, Seng X, Zhang Z (2010) Thermodynamic study on metastable phase: from polycrystalline to nanocrystalline system. Appl Phys Lett 97:181911 (1–3)

21. Xu W, Song X, Lu N et al (2009) Nanoscale thermodynamic study on phase transformation in the nanocrystalline Sm_2Co_{17} alloy. Nanoscale 1:238–244
22. Xu W, Song X, Zhang ZX (2012) Multiphase equilibrium, phase stability and phase transformation in nanocrystalline alloy systems. Nano Brief Rep Rev 7:125012 (1–10)
23. Chookajorn T, Schuh CA (2014) Thermodynamics of stable nanocrystalline alloys. Phys Rev B 89:064102 (1–10)
24. Zhang RF, Veprek S (2008) Phase stability of self-organized nc-TiN/a-Si_3N_4 nanocomposites and of $Ti_{1-x}Si_xN_y$ solid solution studied by *ab initio* calculation and thermodynamic modeling. Thin Solid Films 516:2264–2275
25. Sheng SH, Zhang RF, Veprek S (2011) Phase stabilities and decomposition mechanism in the Zr–Si–N system studied by combined ab initio DFT and thermodynamic calculation. Acta Mater 59:297–307
26. Sheng SH, Zhang RF, Veprek S (2011) Study of spinodal decomposition and formation of nc-Al_2O_3/ZrO_2 nanocomposites by combined ab initio density functional theory and thermodynamic modeling. Acta Mater 59:3498–3509
27. Ivashchenko VI, Veprek S, Turchi PEA et al (2012) Comparative first-principles study of TiN/SiN_x/TiN interfaces, Phys Rev B 85:195403 (1–14)
28. Ivashchenko VI, Veprek S, Turchi PEA et al (2012) First-principles study of TiN/SiC/TiN interfaces in superhard nanocomposites. Phys Rev B 86:014110 (1–8)
29. Ivashchenko VI, Veprek S (2013) First-principles molecular dynamics study of the thermal stability of the BN, AlN, SiC and SiN interfacial layers in TiN-based heterostructures: comparison with experiments. Thin Sol Films 545:391–400
30. Rusanov AI (2002) The surprising world of nanostructures. Russ J Gen Chem 72:495–512
31. Glezer AM (2002) Amorphfnye i nanocrystallicheskie struktury: skhodstvo, razlichiya ivzaimnye perekhody (Amorphous and nanocrystalline structures: similarity, differences, and mutual transitions). Ros Khim Zhurn 46(5):57–64 (in Russian)
32. Ovid'ko IA (2009) Theories of grain growth and methods of its suppression in nano- and polycrystalline materials. Mater Phys Mech 8:174–199
33. Castro RHR (2013) On the thermodynamic stability of nanocrystalline ceramics. Mater Lett 96:45–56
34. Andrievski RA (2014) Review of thermal stability of nanomaterials. J Mater Sci 49:1449–1460
35. Tschopp MA, Murdoch HA, Kecskes LJ et al (2014) "Bulk" nanocrystalline metals: review on the current state of the art and future opportunities for copper and copper alloys. JOM 66:1000–1019
36. Atwater MA, Scattergood RO, Koch CC (2013) The stabilization of nanocrystalline copper by zirconium. Mater Sci Eng A 559:250–256
37. Chookajorn T, Schuh CA (2014) Nanoscale segregation behavior and high-temperature stability of nanocrystalline W–20 at%Ti. Acta Mater 73:128–138
38. Koch CC, Scattergood RO, Saber M et al (2013) High temperature stabilization of nanocrystalline grain size: thermodynamic versus kinetic strategies. J Mater Res 28:1785–1791
39. Atwater MA, Roy D, Darling KA et al (2012) The thermal stability of nanocrystalline copper cryogenically milled with tungsten. Mater Sci Eng A 558:226–233
40. Darling KA, Roberts AJ, Mishin Y et al (2013) Grain size stabilization of nanocrystalline copper at high temperature by alloying with tantalum. J All Comp 573:142–150
41. Frolov T, Darling KA, Kecskes LJ et al (2012) Stabilization and strengthening of nanocrystalline copper by alloying with tantalum. Acta Mater 60:2158–2168
42. Özerinç S, Tai K, Vo NQ et al (2012) Grain boundary doping strengthens nanocrystalline copper alloys. Scr Mater 67:720–723
43. Koch CC, Scattergood RO, VanLeeuwen BK et al (2012) Thermodynamic stabilization of grain size in nanocrystalline metals. Mater Sci Forum 715–716:323–328
44. Darling KA, Kecskes LJ, Atwater M et al (2013) Thermal stability of nanocrystalline nickel with yttrium additions. J Mater Res 28:1813–1819

45. Anderoglu O, Misra A, Wang H et al (2008) Thermal stability of sputtered Cu films with nanoscale growth twins. J Appl Phys 103:094322 (1–6)
46. Lu L, Shen Y, Chen X et al (2004) Ultrahigh strength and high electrical conductivity in copper. Science 304:422–426
47. Liu X, Zhang HW, Lu K (2013) Strain-induced ultrahard and ultrastable nanolaminated structure in nickel. Science 342:337–340
48. Zheng S, Beyerlein IJ, Carpenter JS et al (2013) High-strength and thermally stable bulk nanolayered composites due to twin-induced interfaces. Nature Commun 4:1696–1703
49. Ames M, Markmann J, Karos R et al (2008) Unraveling the nature of room temperature grain growth in nanocrystalline materials. Acta Mater 56:4255–4266
50. Gottstein G, Shvindlerman LS, Zhao B (2010) Thermodynamics and kinetics of grain boundary triple junctions in metals: recent developments. Scr Mater 62:914–917
51. Konkova T, Mironov S, Korznikov A et al (2010) Microstructure instability in cryogenically deformed copper. Scr Mater 63:921–924
52. Cheng L, Hibbard GD (2008) Abnormal grain growth via migration of planar growth interfaces. Mater Sci Eng A 492:128–133
53. Hattar K, Follstaedt DM, Knapp JA et al (2008) Defect structures created during abnormal grain growth in pulsed-laser deposited nickel. Acta Mater 56:794–801
54. Kacher J, Robertson IM, Nowell M et al (2011) Study of rapid grain boundary migration in a nanocrystalline Ni thin film. Mater Sci Eng A 528:1628–1635
55. Paul H, Krill CE III (2011) Abnormally linear grain growth in nanocrystalline Fe. Scr Mater 65:5–8
56. Kotan H, Darling KA, Saber M et al (2013) An in situ experimental study of grain growth in a nanocrystalline $Fe_{91}Ni_6Zr_1$ alloy. J Mater Sci 48:2251–2257
57. Mannesson K, Jeppsson J, Borgenstam A et al (2011) Carbide grain growth in cemented carbides. Acta Mater 59:1912–1923
58. McKie A, Herrmann M, Sigalas I et al (2013) Supresion of abnormal grain growth in fine grained polycrystalline diamond materials (PCD). Int J Refr Met Hard Mater 41:66–72
59. Novikov VY (2011) On abnormal grain growth in nanocrystalline materials induced by small particles. Int J Mater Res 4:446–451
60. Novikov VY (2013) Grain growth suppression in nanocrystalline materials. Mater Lett 100:271–273
61. Franke O, Leisen D, Gleiter H et al (2014) Thermal and plastic behavior of nanoglasses. J Mater Res 29:1210–1216
62. Hultman L, Mitterer C (2006) Thermal stability of advanced nanostructured wear resistant coatings. In: Cavaleiro A, De Hosson JT (eds) Nanostructured coatings. Springer, New York, pp 609–656
63. Mayrhofer PH, Mitterer C, Hultman L et al (2006) Microstructural design of hard coatings. Progr Mater Sci 51:1032–1114
64. Andrievski RA (2007) Nanostructured superhard films as typical nanomaterials. Surf Coat Technol 201:6112–6116
65. Levashov EA, Shtansky DV (2007) Multifunctional nanostructured films. Russ Chem Rev 76:463–470
66. Pogrebnjak AD, Shpak AP, Azarenkov NA et al (2009) Structures and properties of hard and superhard nanocomposite coatings. Phys-Usp 52:29–54
67. Veprek S (2013) Recent search for new superhard materials: Go nano! J Vac Sci Technol A 31:050822 (1–33)
68. Schlögl M, Paulitsch J, Mayrhofer PH (2014) Thermal stability of CrN/AlN superlattice coatings. Surf Coat Technol 240:250–254
69. Kuptsov KA, Kiryukhantsev-Korneev PhV, Sheveyko AN et al (2011) Structural transformation in TiAlSiCN coatings in the temperature range 900–1600 °C. Acta Mater 83:408–418

70. Firstov SA, Gorban' VF, Danilenko NI et al (2014) Thermal stability of ultrahard nitride coatings from high-entropy multicomponent Ti–V–Zr–Nb–Hf alloy. Powd Metall Met Ceram 52(9–10):560–566
71. Shtansky DV, Kuptsov KA, Kiryukhantsev-Korneev PhV et al (2012) High thermal stability of TiAlSiCN coatings with "comb" like nanocrystalline structure. Surf Coat Technol 206:4840–4849
72. Pogrebnjak AD (2013) Structure and properties of nanostructured (Ti–Hf–Zr–V–Nb)N coatings. J Nanomater 2013:780125 (1–12)
73. Veprek S (1999) The search for novel superhard materials. J Vac Sci Technol A Vac Surf Films 17:2401–2420
74. Veprek S, Veprek-Hejman MJG (2008) Industrial applications of superhard nanocomposites coatings. Surf Coat Technol 202:5063–5073
75. Veprek S, Argon AS (2002) Towards the understanding of mechanical properties of super- and ultrahard nanocomposites. J Vac Sci Technol B 20:650–664
76. Männling H-D, Patil DS, Moto K et al (2001) Thermal stability of superhard nanocomposites coatings consisting immiscible nitrides. Surf Coat Technol 146–147:263–267
77. Kiryukhantsev-Korneev PhV, Shtansky DV, Petrzhik MI et al (2007) Thermal stability and oxidation resistance of Ti–B–N, Ti–Cr–B–N, Ti–Si–B–N, and Ti–Al–Si–B–N films. Surf Coat Technol 201:6143–6147
78. Chui P, Sun K (2014) Thermal stability of a nanostructured layer on the surface of 316L stainless steel. J Mater Res 29:556–560
79. Wang Q, Yin Y, Sun Q et al (2014) Gradient nano microstructure and its formation in pure titanium produced by surface rolling treatment. J Mater Res 29:569–577
80. Detor AJ, Deal AD, Hanlon T (2012) Grain boundary engineering alloy 706 for improved high temperature performance. In: Huron ES, Reed RC, Hardy Mc et al (eds), Superalloys 2012: 12th international symposium on superalloys, TMS, Wiley, Hoboken, pp 873–880
81. Saber M, Koch CC, Scattergood RO (2015) Thermodynamic grain size stabilization models: an overview. Mater Rev Lett 3:65–75
82. Ma T, Hedström P, Ström V et al (2015) Self-organizing nanostructured lamellar (Ti, Zr)C—a superhard mixed carbide. Int J Refr Met Hard Mater 51:25–28

Chapter 3
Nanomaterials Behavior under Irradiation Impact

Abstract In this chapter, the main attention is given to the possible effects of irradiation on the structure and properties of NMs. The data on various nanostructures behavior under irradiation by high energy ions/neutrons as well as the possibilities and potentialities of some microscopic approaches and MD modeling results are considered and generalized. The examples of possible practical use of NMs are presented and several poorly investigated problems are discussed.

3.1 General Considerations

A very important role of various materials in the general nuclear industry progress is well known, and it is reasonable to consider the use of namely NMs in this field, especially bearing in mind their stability under irradiation, with accounting the temperature, loadings, corrosive media action, etc. Let us remember a general picture of the origin of radiation defects in a solid body during radiation. As a rule, it is suggested that under irradiation by high energy ions and neutrons, there is observed a partial energy transfer to the displaced atoms of crystalline lattice, which leads to the formation of primary knocked-out atoms (PKA). That process generates so-called displacement cascades containing the Frenkel pairs in the form of both interstitial atoms and vacancies (IAV) and their complexes (clusters) in the form of vacancy loops (or vacancy nanopores) as well as dislocations. The point radiation defects can collide (and annihilate) both at meeting each other and/or after moving further apart along GBs, which in such a situation act as sinks for radiation defects. An intense irradiation of material leads to the following general effects influencing the material structure, content, and properties: material amorphization; plasticity decrease or embrittlement (this effect has been long known in material science and called as radiation hardening); high-temperature creep growth; intensification of boundary segregation processes; material swelling as a result of the non-compensated vacancy and atomic sinks; surface blistering under ion irradiation, and elements transmutation owing to some nuclear reactions and decays.

© Springer International Publishing Switzerland 2016
R.A. Andrievski and A.V. Khatchoyan, *Nanomaterials in Extreme Environments*,
Springer Series in Materials Science 230, DOI 10.1007/978-3-319-25331-2_3

Compared to traditional CG materials (e.g., [1]), information, concerning the generation of radiation defects, their development and properties in various NMs, is very poor since many of these studies are still in their infancy (e.g., [2–7]). It is obvious that the presence of the developed interface system in NMs must increase its role as radiation defects sinks (as compared to the CG counterparts), and so the negative action of radiation effects can be attenuated. However, on the other hand, such accumulation of radiation defects at GBs can initiate nanostructure amorphization processes during which the displacement cascade development has its own specificity in macro- and nanostructures. It is important that in such situation, we cannot unambiguously estimate the positive and negative influences of the irradiation action. From the given general considerations and the available data, at least, four different versions of the NMs behavior under irradiation can be distinguished:

(1) In the NMs with a developed system of various interfaces (such as GBs and TJs) performing the role of radiation defect sinks, the radiation resistance increase is observed in comparison with their traditional CG counterparts.
(2) In some NMs, irradiation can promote annihilation of nanostructures and thus leads to the transformation of them into an amorphous state.
(3) The presence of radiation defects can generate or stimulate various recrystallization processes, which can influence the two above described versions of material transformation.
(4) Finally, transmutation processes should be taken into account, especially in the case of irradiated fuels and boron-containing materials (for example, in the latter case, due to (n, α) reactions the materials are depleted of boron and demonstrate swelling caused by helium accumulation in pores).

Besides, one must distinct the complicated peculiarities of the behavior of various samples under irradiation (bulk specimens and thin films, objects with metallic and covalent bonds, etc.) as well as different penetration ability of ions and neutrons in materials under study. Thus, the problem the NMs behavior under irradiation requires the consideration of many factors.

3.2 Main Experimental Results

3.2.1 Ion Irradiation

The ion irradiation in accelerators cannot trace all the features of radiation defects behavior, but this approach is widely used in the current experimental practice thanks to its economic and operative advantages in comparison with the so-called reactor experiments. Investigations of the effect of ion implantation on the material properties started in the 80th of last century (see survey [8]), but systematic studies in the NMs field (especially with accounting the GS effect) have been carried out

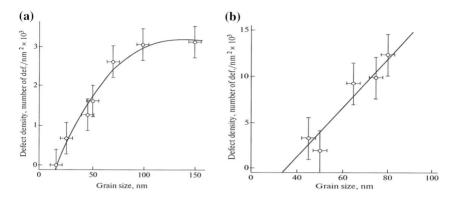

Fig. 3.1 The GS effect on the radiation defect density in zirconium oxide samples (**a**) and palladium those (**b**) irradiated with Kr ions at 293 K (ZrO$_2$: energy $E = 4$ MeV according to dose of 3–8 displacements pear atom (dpa); Pd: $E = 0.24$ MeV according to dose of 110–210 dpa) (adapted from [2])

only over about 20 years, and the first such results were presented at the Second International Nanomaterials Conference NANO1994 in Stuttgart [2, 9]. Figure 3.1 shows the radiation defect density changes due to irradiation of ZrO$_2$ and Pd samples. One can see that there is a marked decrease in the radiation defect density for the sizes ranging from 100–150 to 20–40 nm. At a grain size below 15 nm (ZrO$_2$) and 30 nm (Pd), defects in irradiated samples were not observed at all [2, 9], which indicates an elevated radiation stability of nanostructures.

By now, other authors have confirmed the conclusions [2, 9]. Some characteristic and interesting results obtained over 2004–2014 are presented in Table 3.1 with short description of methods for sample preparation and the irradiation conditions. The presented data convincingly indicate that practically in all studied materials, such as metals, alloys, steels, intermetallic compounds, and oxides, prepared by various nanotechnology methods, the nanostructures turned out to be more resistant to ion irradiation compared to the microstructures in their traditional CG counterparts.

In addition, it should be marked that numerous and demonstrative results, which were obtained in experiments with multi-layer films, can be added to Table 3.1. The interlayers in such systems serve as sinks radiation defects. Using layers with non-miscible components and varying their individual thicknesses so that the summed film thickness can remain constant, one can easily trace the influence of the number of individual layers on the material microhardness, swelling and other characteristics under irradiation. An illustrative example is presented in Fig. 3.2, where changes in the radiation hardening (a) and swelling (b) for some multi-layer Cu/V films are presented as functions of the reciprocal values of the individual layer thicknesses. The results descriptively show that the thinner individual layers in a film (i.e., the greater the number of interfaces acting as sinks for radiation defects), the smaller the increase in its hardness and helium pores formation.

Table 3.1 Effect of ion irradiation on NMs

Materials and manufacturing method	Grain size L (nm)	Conditions of irradiation			T K	Main result
		Ion	E (MeV)	Dose Fluence (dpa) (ion/m²)		
Ni (electrodeposition)	20–30	Ni	0.84	5	293	No change in L was registered; the SFT formation was observed
Ni (HPT)	115	Proton	0.59	0.56	293	L decreased to 38 nm; SFT and twinned boundaries (TBs) were formed; hardness increased in about 2 times
Cu-0.5Al₂O₃(HPT) [10, 11]	~180	Proton	0.59	0.91	293	L increased to 495 nm; SFT are formed; hardness remained unaltered. In all cases, the radiation- induced SFT density in NMs was smaller than that in CG materials
MgCa₂O₄ (consolidation at 5 GPa) [12]	4–12	Kr	0.3	12–96	100	Amorphization was not observed. Amorphization started at 12 dpa (at L >1 μm)
TiNi (HPT) [13]	31 ± 6	Ar	1.5	0.2–5.6	293	Amorphization was not observed. Amorphization started at 0.4–2.5 dpa (at L >100 μm)
W-0.5TiC (hot isostatic pressing (HIP) [14]	50–200	He	3	2 × 10²³	823	Critical fluence for blistering is tenfold that of CG samples
UFG-316SS austenitic steel (HPT) [15]	40	Fe	0.16 / 10	10 / 10	623	Radiation-induced segregations at GBs are smaller in nanosteel
Fe film (magnetron sputtering) [16]	49–96	He	0.1	6 × 10²⁰	293	The density of He bubbles and radiation hardening are smaller in nanofilms
14YWT ferritic steel (hot extrusion and rolling HE/R) [17]	L = 200–400; nano-clusters of 1–5	Pt	10	15–160	173–1023	The main beneficial features of nanostructure remain almost unchanged after irradiation at all studied conditions

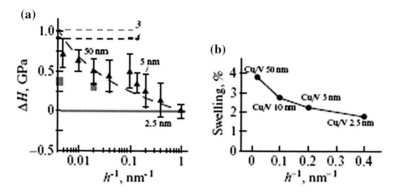

Fig. 3.2 The radiation hardening (a) and swelling (b) as a function of the reciprocal thickness of one layer in Cu/V multi-layer films irradiated with He ions ($E = 50$ keV; fluence of 6×10^{20} m^{-2}; $T = 20$ °C). Lines 3 and 4 (a) relate to the ΔH for Cu and V individual films, correspondingly [18]

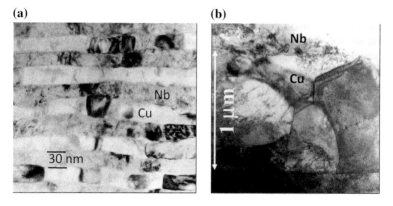

Fig. 3.3 TEM images of rolled multi-layer Cu/Nb samples with different one layer thickness (h) after an irradiation with He ions ($E = 0.15$ MeV; fluence of 10^{21} m^{-2}; $T = 20$ °C): **a** $h = 30$ nm; **b** $h = 1$ μm (adapted from [19])

In general, the presence of thin individual layers promotes preservation of a layered morphology after irradiation, which is conveniently illustrated for the Cu/Nb films in Fig. 3.3, where in the case of thin individual layers (a) a nanolaminated structure remains after radiation without any evidence to the presence of dislocation defects, which radically differ from the thick-layer system behavior (b), where we see characteristic disorders of the layers, dislocation-cell structures, and even some marks of grain rotation. Analogous results demonstrating a similar role of such thin individual layers under irradiation were obtained by many authors for other various multi-layer films (Cu/W, W/ZrO$_2$, Ta/Ti, Ag/Ni, CrN/AlTiN, etc. [20–24]), where in some cases the interfaces are not purely metallic but mixed (metal-oxide) and nitride ones.

A more complicated situation is observed in the case of irradiation of the nanostructured wide-zone semiconductor β-SiC (3C-SiC), a very interesting material not only for nuclear industry, but also for nanoelectronics, chemical technology, and some biomedicine applications [25, 26]. It is known from literature that the β-SiC nanostructure is tolerant to the irradiation with 4 MeV Au ions [27] and, in contrary, the behavior of irradiated nanostructured samples (L = 30–50 nm) and single crystals looks much the same [28, 29]. However, a more detailed following study of nanostructured β-SiC films and single crystals, irradiated with 0.55 MeV Si ions, has shown that at room temperatures the total amorphization of these objects occurs at irradiation dozes of ~ 3 and ~ 0.29 dpa, correspondingly [30]. Such preferential nanostructure tolerance is connected with a high density of planar defects (in the form of stacking faults), which promote the IAV recombination processes leading to self-healing of the generated radiation defects. The following TEM in situ examination of electron irradiated silicon carbide films confirmed the important role of planar defects acting namely as sinks and traps in the IAV processes [31].

The features of irradiation induced amorphization in some nanostructures can be illustrated by the results of the nanoparticles behavior studies in inert SiO_2 matrices presented in Table 3.2.

Particularly representative results were obtained for zirconium oxide: ZrO_2 nanoparticles amorphizate under the action of a moderate irradiation doze with Xe ions, whereas the oxide single crystal preserves its crystal state even at high irradiation dozes. As shown in Table 3.2, the dimension-size effects are observed for Cu and Ge, but Au nanoparticles do not amorphizate in the studied size range. The

Table 3.2 Irradiation effect on nanoparticles embedded in amorphous silica (SiO_2)

Subject	Size of nanocrystal (nm)	Irradiation conditions			Main result
		Ion	E (MeV)	Dose (dpa) Fluence (ions/m²)	
ZrO_2/SiO_2 [32] ZrO_2 [33]	~ 3	Xe	1	~ 0.8	Amorphization
	Single crystal	Xe	0.4	680	Preservation of crystal state
Cu/SiO_2 [34, 35]	~ 2.5	Sn	5	0.16	Amorphization
	~ 8	Sn	5	10^{19}	Preservation of crystal state
Au/SiO_2 [32, 36]	3	Xe	1	~ 0.8	Preservation of crystal state
	3–5	Sn	2.3	10^{19-20}	Preservation of crystal state
Ge/SiO_2 [37, 38]	4–8	Si	5	10^{15-19}	Amorphization of nanocrystals is observed firstly
Co/SiO_2 [39]	3.7 ± 1.0	Au	9	10^{17}	Amorphization

problem of the nanoparticles behavior is complicated, and readers can find more detailed information in review [40]. Especially for the oxides irradiation by fast neutrons, it was shown that a spontaneous radiation amorphization starts at some critical defect concentration when the ions displacement ranges up to the value known as the Lindemann melting criterion [41], but in general, the problems of size effects in such systems remain to be solved.

It must be marked that the studies of the ion irradiation influence on the NMs structure and properties offer contradicted experimental results, and thus scientists come up against complex and non-ordinary problems requiring modern techniques or combination of the available ones. Now, in the field, there are many diverse and independent modern methods, such as TEM (including high-resolution variant HRTEM), scanning electron microscopy (SEM including high-resolution variant HRSEM), XRD methods, including small-angle X-ray scattering, glancing-incident angle X-ray diffraction (GIAXRD), extended X-ray absorption fine structure spectroscopy, X-ray absorption near-edge spectroscopy, energy dispersive spectroscopy (EDS), selected area electron diffraction (SAED), Rutherford backscattering spectrometry, Auger electron spectroscopy, electron backscattering diffraction (EBSD), atom probe tomography (APT), differential thermal analysis/thermal gravimetric analysis (DTA/TGA), atomic force microscopy, Raman spectroscopy as well as improved or advanced micro/nanoindentation tests and electrical/magnetic measurements. For theoretical considerations and estimations, there were elaborated some new modeling methods, and now ion penetration and irradiation damage profiles can be calculated with the use of the known software stopping and range of ions in matter (SRIM) program. As an example, the calculated damage and implantation profiles are presented in Fig. 3.4 for β-SiC samples (with GS of 30–50 nm) irradiated with 4 MeV Au ions (adapted from [29]).

Fig. 3.4 The damage profile in SiC (*continuous line*) and Au implantation profile (*dotted line*)

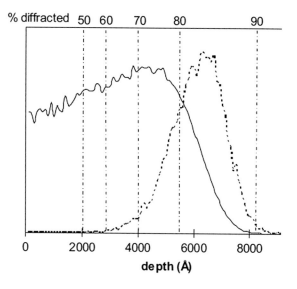

On the upper line of Fig. 3.4, the depth of the diffraction measurements by GIAXRD is presented in %, and the results allow one to make a conclusion that about 80 % of the irradiation depth can be connected with the damage region and about 20 % relates to the mixed zone (implantation + damage). Such estimations can be important for analysis of changes in the NMs properties under an ion irradiation.

Besides the above presented data on the NMs tolerance to irradiation and their possible amorphization, there are some other listed below interesting results waiting for consideration and interpretation.

1. The studies of the GG in nanostructured metallic films (Au, Pt, Cu, Zr, and Zr–Fe), exposed to irradiation with Ar ($E = 0.5$ MeV) and Kr ($E = 0.5$–1 MeV) ions [42], have shown that there are three distinct ranges related to temperature: a purely thermal one (where temperature is a dominant factor of GG), thermally induced or mixed (where thermal and radiation effects are summarized), and low-temperature one (where the thermal effect on GG is very weak). The temperatures of transitions from mixed to low-temperature ranges depend on the object under study, but they oscillate by the homological temperature scale between 0.15 and 0.20 T_m. Based on these data, a theory of radiation-induced GG was developed supposing some thermal peaks formation under irradiation in cascades and subcascades. In the frame of this theory, the radiation-induced grain growth can be described by the expression

$$L^n - L_o^n \sim K\Phi t, \tag{3.1}$$

where $n \sim 3$, Φ is the ion beam intensity (ion/m^2 s), t is the time, K is a constant dependent on the grain mobility and moving force, L and L_o correspond to the moving and initial values of the GS.

2. For nickel and copper, the growth of the radiation defects number with increasing the irradiation was detected especially for samples having the SFT (with a mean size of 2.5–4.4 nm) and TB forms [10, 11]. It is interesting that in Ni, this dependence is accompanied by decrease in GS from 115 nm down to 38 nm, whereas by contrast in copper, the size growth from 178 to 493 nm was observed. The authors [11] especially marked this fact and underlined that such distinction needs the further studies and considerations. The mean GS increase from 40 to 60 nm was also observed in nanostructured austenitic steel 316SS irradiated with 0.16 MeV Fe ions [43]. The sufficient difference in the nanograin growth rates was shown [44] for the tetragonal and cubic ZrO_2 crystal modifications irradiated with Kr ions ($E = 0.34$ MeV).

3. The radiation induced processes and their role in the GBs enrichment/depletion with different elements (as well as the chromic steels reinforcement by some nanoinclusions) were widely and fruitfully investigated in several works, where the combination of TEM, EBSD and APT methods was used (e.g., [45, 46]). As an example, Fig. 3.5 demonstrates the evolution of oxide nanoinclusions

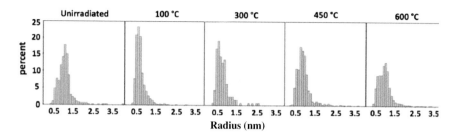

Fig. 3.5 TEM data concerning the irradiation temperature effect on the oxide nanoinclusion distribution histograms after irradiation of 14 YWT steel with Ni ions (adapted from [46])

(Fe, Ti, Y)O_X for the 14YWT steel samples irradiated with Ni ions ($E = 5$ MeV, dose of 100 dpa). From the other side, there were obtained element distribution maps for some nanostructures [46]. Naturally, such data are important for estimation of changes in the mechanical and corrosion properties of widely used alloyed steels after irradiation.

4. The high-temperature strength and creep of copper and its alloys under irradiation with Kr ions ($E = 1.8$ MeV, dose up to 75–100 dpa) were studied by Averback et al. [47, 48], and it was found that a marked growth of high-temperature inclusions begins at temperatures ~ 0.65 and ~ 0.85 T_m (corresponding to the Mo and W additions). The dependence of the creep velocity on the stress and irradiation dose proved to be linear.

5. In porous samples, the open pores surface rise is favoring to the removal of radiation defects, i.e., if the sinks number growth is increasing, then such porous objects increases their radiation stability as compared with usual dense samples. This phenomena was been clearly demonstrated for the irradiated porous samples of ZrO_2 [49] and Au [50].

6. In the works of Yu, Zhang et al. [24, 51–53], the methods of HRTEM (including in situ studies of irradiation-induced objects), XRD, and SAED revealed the TBs properties as sinks of radiation defects. The studies of interfaces in multi-layer films of Ag/Ni, Ag/Al, Cu/Fe, and Cu/Ni types allow formulating the following two main criteria of the TBs formation in the described FCC metal structures: a low value of the TBs energy and their high coherence.

7. In Fig. 3.6, radiation damages are compared for single crystal (SC), CG ($L = 20$–30 μm), and nanocrystal grain (NG) copper samples irradiated with He ions. The results show marked distinctions in both the number and sizes of radiation pores in samples with different GS. At the same time, it was detected that in single crystal samples (with initial mean GS of ~ 15 nm) the radiation-induced defects are smaller and preferentially oriented along the GBs (the mean GS in irradiated NG samples was ~ 35 nm).

8. In nanocrystalline W films under irradiation by Au ions ($E = 4$ MeV; dose range from 6 to 100 dpa) at radiation doses over 30 dpa, the transformation of the primitive cubic phase (β) into the BCC modification (α-phase) was observed [55].

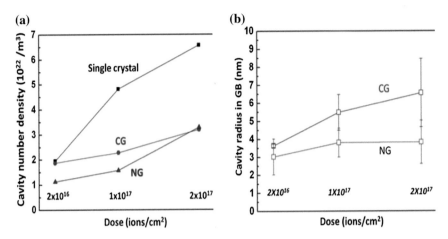

Fig. 3.6 The dose effect on the cavity number density (**a**) and cavity radius (**b**) in single crystal, CG, and NG Cu samples after irradiation with He ions at 450 °C (adapted from [54])

From the other side, the irradiation leads to the GS growth from ~ 13 to ~ 19 nm, which corresponds to the above mentioned expression (3.1) which was put forward in [42].

3.2.2 Neutron Irradiation

The positive role of nanostructures in the material radiation stability growth, revealed in the experiments with ion irradiation [2, 3, 9–13], on the one hand, initiated the reactor tests of various NMs and, on the other hand, became a base for some new nanostructured steels development, specially designed to be used in fission and fusion reactors of the IV generation. Table 3.3 lists the main results concerning the considered NMs structures and the properties changes initiated by neutron irradiation.

These data make it is obvious that a GS decrease practically always raises the material radiation stability, just as it was observed in the ion irradiation tests (see Table 3.1). At the same time, it should be marked that the number of reactor tests and their duration remain limited and general information concerning the behavior of materials with small nanograins (below 100 nm) under the real conditions of reactor operational regimes seems to be absent. The problems are connected with both the difficulties of the reactor tests themselves and the deficiency of the required bulk NMs.

The effectiveness and advantages of the described radiation studies in general can be illustrated by the presented in Table 3.4 results obtained in comparative tests carried out especially for two types of the structural steels, namely, the CG steel ODS-EUROFER (developed in West Europe) and new nanostructured steel

Table 3.3 Effect of neutron irradiation on NMs and some materials with related structure (oxide dispersion-strengthened (ODS) steel (CVD means chemical vapor deposition; CVI does chemically vapor-infiltrated)

Material (manufacturing)	Grain size (nm)	Conditions of irradiation			Main result
		E (MeV)	Dose (dpa) Fluence (n/m²)	T (K)	
Modified ferritic/martensitic ODS 9–12Cr steels (HE/R) [56]	10^3–10^4(grains); ~4 (inclusions)	>0.1	15	670–807	The size and content of nanoinclusions (near $Y_2Ti_2O_7$) retain stability
UFG SUS316L + 1 %TiC (HIP and rolling); Ni (electrodeposition); Ni–W (the same) [57]	100–300; 650	>1	10^{23}–10^{24}	560	Small or no increase hardness was observed after irradiation
W–0.5TiC (HIP) [14]	50–200	>1	2×10^{24}	873	Radiation hardening and defects were smaller in nano W
Austenitic UFG-CW 316SS (HPT) [58]	40; 40×10^3	>1	12	633	Decreasing GS results in more precise the segregation study
Nanostructured ferritic steel 14YWT (HE/R) [59–61]	100–300 (grains); 1–4 (inclusions)	>0.1	1.2–1.6	573–973	Small radiation hardening and plasticity some decrease
Nanostructured ferriticsteel MA957 (HE/R) [62]	1200/300 (ℓ/L)	>0.1	3	873	The size (~2 nm) and content of nanoinclusions (Y–Ti–O) retain stability
SiC and its composites SiC/SiC (CVD and CVI) [63]	10^4–10^5	>0.1	~28 40.7	573, 923 1073	Swelling, thermal conductivity and strength have been studied in detail. Irradiation-insensitivity was demonstrated
Austenitic AISI 321 steel (ECAP) [64]	300–400; 40×10^3	>1	5.3	293–623	Radiation hardening and plasticity decrease are smaller after ECAP
UFG low carbon steel (ECAP) [65]	370 ± 60; ~44×10^3	>1	1.15×10^{-3}	328	Irradiation hardening and ductility decrease were smaller after ECAP
Creep of nanopowder sintered SiC (sintering additions and hot pressing) [66]	216, 414	>0.1	1.9	653–1453	The effect of GS and irradiation was studied and discussed

Table 3.4 Yield strength (σ_Y), ultimate tensile strength (σ_{UTS}), total elongation (δ), fracture toughness (K_{IC}), temperature of brittle to ductile transition (T_{BDT}), and difference of T_{BDT} (ΔT_{BDT}) for ODS-EUROFER and 14YWT steels in initial and neutron-irradiated ($E > 0.1$ MeV, dose of 1.5 dpa, and $T = 300$ °C) states

Steel	State	σ_Y (MPa)	σ_{UTS} (MPa)	δ (%)	K_{IC} (MPa m$^{0.5}$)	T_{BDT} (°C)	ΔT_{BDT}
ODS-EUROFER	Initial	966	1085	11.7	160	−115	85
	Irradiated	1243	1254	7.1	180	−30	
14YWT	Initial	1435	1564	12.0	175	−188	12
	Irradiated	1560	1641	7.4	225	−176	

14YWT, offered by Oak Ridge National Laboratory (USA) and developed on the modern approaches in the field [60]. These results testify the higher mechanical properties of the 14YWT steel in both the initial and irradiated states. Of special interest are the facts of sufficiently less radiation hardening and the temperature shift (ΔT_{BDT}) in irradiated samples for the transition from a brittle state to a ductile one. These advantages must be connected namely with a 14YWT steel structure, including the elongated grains (with length to 1–1.5 μm and mean width ∼ 300 nm) and oxide inclusions with sizes of ∼ 1–2 nm (see Fig. 3.5), that is markedly less than the ODS-EUROFER steel structural parameters. Besides, the nanoinclusions density in 14YWT steel is ∼ 2 × 10^{24} m^{-3} and their characteristic features are a high coherence with the matrix phase and lower probability of the concentration stresses occurrence and, as a corresponding result, the nanocracks generation.

The thermal conductivity and swelling evolutions in SiC fibers, irradiated with fast neutrons ($E > 0.1$ MeV), are presented in Fig. 3.7, and it is easy to see that curves, describing two these characteristics behavior (a thermal conductivity decrease and a swelling increase) are stabilizing at an irradiation dose about 0.1–1 dpa. This level is wholly satisfactory one, and the studied materials (SiC fibers and impregnated SiC/SiC composites) can be considered as the radiation-resistive materials at least in the studied ranges of the neutron irradiation conditions (doses of 30–40 dpa, $T = 300–800$ °C). It should be marked that under these conditions the material mechanical properties are scarcely affected by irradiation, and the effect manifests itself only in the increase statistical spread of strength parameters, that is reflecting in a general characters of the so-called Weibull's relationships.

It is worthy to mention that in more complex systems the dependence of the material properties on the irradiation dose can have a more complicated character. For example, the studies of SiC samples sintered from the nanopowders with some activating additives have shown that irradiation elevates the creep velocity as well, but the GS effect manifests itself only in the 226–414 nm GS interval and at low radiation dozes (0.011–0.11 dpa) [66]. For the fibers with CG structure ($L = 5–10$ μm), irradiation also elevates the creep velocity, but herein the GS effect is not observed in the 380–540 °C range and becomes detectable only at $T = 760–790$ °C. The radiation creep of the sintered SiC samples in the described condition is about 1.4–1.8 times greater than that for the fibers with a CG structure.

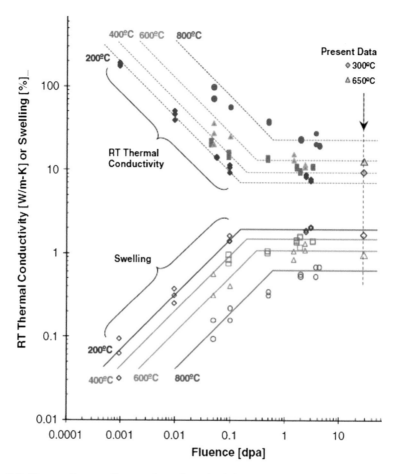

Fig. 3.7 Neutron fluence effect on thermal conductivity and swelling of CVD SiC (adapted from [63])

3.3   Some Theoretical Approaches and Modeling

A theoretical model describing a radiation-induced amorphization process in nanostructured objects was proposed by Ovid'ko and Sheinerman [67]. In this model, the Frenkel pairs IAV generation was suggested in two versions, namely for the high- and low-energy interactions separately. In the former case, the IAV pairs arise both at nano GBs and inside grains, whereas in the latter case, the vacancies come into being only at the GBs but interstitial atoms arise in nanograins. The evolution of radiation defects was theoretically analyzed for the following main stages: (1) generation of radiation-induced defects; (2) absorption of the defects by interfaces; (3) annihilation of IAV pairs; (4) formation of some stable clusters from point defects. For the high-energy interactions, stages (1) and (2) were considered as dominating ones. Herein, a preferential amorphization region (dependent on the nano GS) was defined on the basis of the energy consideration, where the sum of the energies of interface surfaces and point defects elastic energies exceeds the energy of a characteristic threshold for the crystal-to-amorphous state transition. The developed interface net, on the one hand, assists the system free energy increase (i.e., decreases the energetic barrier of amorphization), but, on the other hand, favors to the radiation defects release, i.e., acts against the amorphization and enhances the NMs radiation stability.

A similar energy-based approach was developed by Shen [68], where the author proposed a qualitative model for the ΔG change depending on the GS and made an important conclusion that for every material there is an optimal interval of the GS, where radiation defects release and consequently amorphization resistance can be provided. This approach is schematically presented in Fig. 3.8 (adapted from [68]), where five energetic zones with different processes and characteristics may be defined:

1. Transition to an amorphous state is possible without any irradiation $(L < L_1)$;
2. Transition to an amorphous state is initiated by a weak irradiation $(L_1 < L < L_2)$;
3. Irradiation cannot lead to amorphization $(L_2 < L < L_3)$;
4. Irradiation leads to amorphization $(L_3 < L < L_4)$;
5. In this interval, where the GS exceeds L_4, the ΔG_{pd} value becomes more and more prevailing as with GS growth the interfaces reduce and the role of inter-grain boundaries in the radiation defects release falls off. At some L_m value, the annihilation of defects by the volume recombination mechanism becomes predominant.

The processes in zones 1, 4 and 5 are qualitatively confirmed by the experimental data (e.g., see Fig. 3.1). Some peculiarities of the radiation defects formation in nanocrystals imbedded into a solid matrix were studied by Oksengendler et al. [69].

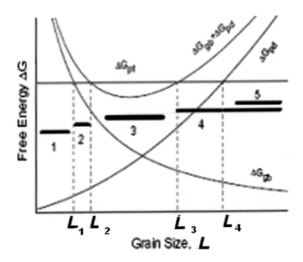

Fig. 3.8 Scheme of the GS effect on the free GB energy (ΔG_{gb}), free energy of point defect (ΔG_{pd}), their sum $(\Delta G_{gb} + \Delta G_{pd})$, and crystal-to-amorphous phase transition (ΔG_{pt}). The designations of regions (1–5) see in text

It was shown that the crystal nanoclusters amorphization process in the inert matrices can be either accelerated (radiation damage case) or slowed-down (radiation resistance case), and the realization of any case depends namely on the situation at the nanocluster–matrix interface: as long as the compressive stresses prevail, the nanoparticles amorphizate, but under the extending stresses the defects formation becomes slow and the crystal material state still preserves. This tendency is qualitatively confirmed by the Table 3.2 data. However, in general, it is difficult to give some proper predictions basing on the results [67–69], because the considered models are characterized by many suppositions and the relevant literature information is absent.

The NMs behavior under irradiation was widely studied using MD methods (e.g., [30, 38–40, 70–81]); the following situations arising in nanostructures under bombardment with ions, neutrons and electrons have been widely described in literature.

1. Amorphization of irradiated nanocrystals in various amorphous matrices [38–40].
2. Generation of displacement cascades in Ni nanograins ($L = 5$ and 12 nm) irradiated with particles with energy of 5–30 keV [70].
3. Radiation-stimulated grain growth in nano-Ni ($L = 5$ and 10 nm) for cascade exited by particle impact with energy of 5 keV [71].
4. Unusual behavior of the interstitial atoms and SFT in damage cascades [72].
5. Evolution of nanocrystal film morphology [73].
6. Unusual behavior of NMs with BCC and FCC structure [74].
7. Unusual behavior of GBs in irradiated nanostructures as defect sinks and sources [75].
8. Generation and growth of so-called He bubbles at interfaces [76].

In these investigations, many important features of the behavior of irradiated nanostructures were exhibited, in particular, such as the main role of intergrain boundaries and TJs acting as radiation defect sinks as well as the role of the radiation-stimulated grain growth processes. Besides the nanofilm stressed state characteristics, changes in the surface roughness were studied and some interesting peculiarities were discovered (for example, in experiments with irradiated BCC and FCC structures, predominance of vacancy clusters is observed in the looser BCC structures).

From the general considerations, it is obvious that defect generation in SiC must be more complicated than in metals because of the presence of two types of mobile interstitial atoms (Si and C) and two types of low-mobile vacancies or their clusters. This situation is very attractive for various MD calculations (e.g., [30, 77–79]), because this method allows one to study the generation and growth of clusters in β-SiC, the role of GS in the process, the behavior of cascades in such systems, etc. In particular, it was strictly shown that the nanolayered stacking faults presence leads to an enhanced mobility of interstitial Si atoms [30].

A vivid example of MD application is given in Fig. 3.9, where a situation is presented in an intergrain boundary for a moment of cascade generation and the radiation defects partially release from the boundary [70]. At the PKA energy

Fig. 3.9 Cross section of a
grain (size of 12 nm)
containing a 5 keV cascade
after the PKA introduction
(**a**) and scheme of GBs acting
as interstitials with free
volume (**b**). The *inset* shows a
magnified image of the defect
region after cooling (adapted
from [70])

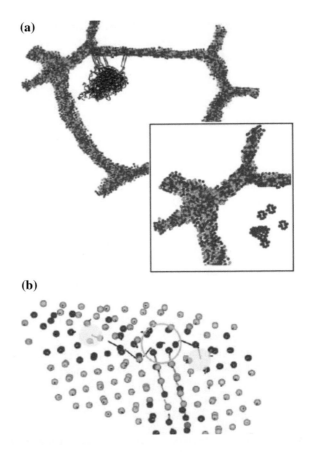

(a)

(b)

elevation, there was fixed the dislocation loops generation, and it is of interest that a
process general picture is sufficiently different, when the experimental results are
compared for the single crystal and nanostructured Ni samples.

In order to properly understand the effectiveness and possibilities of the MD
method, let us consider in detail some results of modeling cascade formation in
vanadium [80] and copper [81, 82]. The crystallites in the modeled BCC vanadium
structures contain tilt and symmetric boundaries of $\Sigma 13$ <320> [001] and $\Sigma 17$
<410> [001] types. The number of atoms in a «calculation cell» equals 65,000–
450,000 depending on the first knocked-on PKA energy in the ranges of <0.5 keV
and >0.5 keV at $T = 10$ K [80]. The cascade evolution was modeled depending on
its duration (more strictly, on its lifetime) for three different stages. The first one is
ballistic (when the energy transferred by PKA spreads throughout the whole model
object volume, and as a result the defects number achieves a maximum and thermal
peaks appear). The second stage is a recombinant one (when the defects number
decreases to some stable value) and the third stage is a diffusion one (further
decrease in the number of defects limited to their interaction and transport processes). The main results of modeling are as follows: the presence of the large

interfaces has a pronounced effect on the development of displacement cascades; in general, they retard these processes due to accumulation of sufficient part of the radiation defects. Sometimes (depending on the PKA energy), the developed interface system can become an impenetrable barrier for a cascade development.

On the other hand, model experiments for the Frenkel pairs IAV interaction with the GBs in FCC copper allowed suggesting another mechanism of such defects annihilation, namely at first boundaries are saturated with high-mobile interstitial atoms, but then their inverse emission and absorption by the vacancies in near-boundary zones take place [81]. For a model system, presenting a combination of the special symmetric tilt boundaries of the $\Sigma 11 <110> \{131\}$ type with the total number of copper atoms in the system about 160.000 (but the number of moving atoms is of $\sim 130,000$), the authors have simulated situations for fifteen cascades with the PKA energy of 4 keV. For the considered copper atoms system, the temperature influence on the IAV release time as well as on the inverse interstitial atoms emission was estimated. The calculation showed that within 10–15 K, the process duration is very great ($t > 10^{10}$ s!) and thus none of the proposed mechanisms works, whereas in the 70–100 K range ($t < 1$ s) the interstitial atoms play a main role and at $T = 300$ K all mechanisms work. Modeling of the concurrent processes of the radiation defects absorption by GBs and their volume recombination has shown that in the case of cascadeless copper irradiation with electrons, the vacancies accumulation from Frenkel pairs IAV on the GBs is under progress (here, the GS decreases from 40 to 15 nm) only during short time intervals, after which absorption of vacancies by coarse grains becomes a prevailing process again [82].

The complicated role of GBs in the radiation-induced defects formation and release for ion compounds was revealed in [83], where in the frame of a multi-scale MD method, the defects generation was studied near the symmetric tilt boundaries in TiO_2 at temperatures of 300 and 1000 K. On the one hand, the importance and possibilities of the MD modeling results must not be overestimated, because these methods often allow tracing only very short-lived processes, leaving many kinetic process peculiarities beyond the consideration. However, on the other hand, it is worthy to mention that the above presented results of computer simulation [81], demonstrating a non-monotonic (by temperature) character of the radiation defects release from NMs, can help to explain, at least qualitatively, some experimental data concerning the electrical resistivity of the Au foil specimens under irradiation with 60 MeV carbon ions at temperatures of 15 and 300 K (Fig. 3.10) [84].

Judging from the electrical resistivity increase, as indicated in Fig. 3.10, the nanostructured samples ($L = 23$ nm) are more sensitive to irradiation at low temperatures in comparison with their CG counterparts, i.e., in situation of the absence of radiation defect release by some above mentioned mechanisms [81]. By contrast, at room temperatures, when all of three radiation defect release mechanisms act, the nanostructured objects demonstrate a higher radiation resistance.

In general, the problems of the displacement cascades development, overlapping, interaction with interfaces, as well as their role in the radiation-induced defects release in NMs require further serious theoretical and experimental investigations.

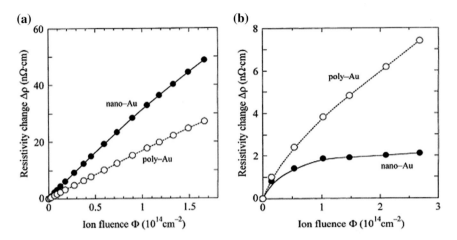

Fig. 3.10 Effect of ion fluence on the electrical resistivity increase of nano–Au ($L = 23$ nm) and poly–Au (annealed at 973 K for ~ 1 h) specimens irradiated with 60 MeV carbon ions at 15 K (**a**) and 300 K (**b**) (accepted from [84])

3.4 Examples of Applications

The main conceptions for the development of the materials, specially adapted to the new IV generation of nuclear technique, were considered in Zinkle's and Was's survey [85], where main demands were formulated for fuel, cladding, and in-core structural materials as applied to the conditions of reactor types, such as fast reactors with different liquid metal or gas cooling (such as sodium fast, lead fast, and gas fast reactors), superhigh-temperature reactors, molten salt reactors, and supercritical-water-cooled reactors with the radiation dozes up to 150–200 dpa. The proposed numerous and various demands have aroused rapid comments of the USA Los Alamos National Laboratory and Massachusetts Institute of Technology specialists [76], who underlined the NMs high tolerance to the irradiation conditions (see Tables 3.1, 3.2 and 3.3) and marked such materials perspectives, namely in connection with the conception demands [85]. First and foremost the problem is connected with nanostructured steels of the 14YWT, MA957, F95, M93, Fe–9/14/18Cr, UFG316SS, and SUS316L types developed in USA, Japan, France, and Russia (Table 3.5).

According to the matrix structure characteristics, most of these steels except for UFG316SS are only adjacent to the NMs dimension high limit ($L \sim 100$ nm), that is, a conventional value is hard to realize in practice correctly for many technological reasons. They mainly belong to the non-nickel specification of the ferritic/martensitic steel class and differ from the classic stainless steels by their low swelling at high irradiation dozes and weak activation after radiation (this factor is important for many operative post-reactor studies).

A short description of technological procedures, connected with production of these steels, is presented in Tables 3.1, 3.2 and 3.3, but a total list of procedures is

Table 3.5 Composition and characteristics of some nanostructured steels

Labeling	Type	Composition (wt%)			L_{grain} (nm)	L_{incl} (nm)
		Cr	Ni	Other elements		
14YWT [17]	Ferritic	14	–	3 W, 0.4Ti, 0.25Y_2O_3	200–400; $l = (1–5)$ 10^3	1–5
MA957 [62]	The same	14.3	–	0.3Mo, 0.9Ti, 0.25Y_2O_3	n/d	1–10
F95[56, 86]	The same	12	–	2 W, 0.3Ti, 0.25Y_2O_3	n/d	~4
M93 [56]	Ferritic/martensitic	9	–	2 W, 0.2Ti, 0.35Y_2O_3	n/d	~4
Fe–9/14/18 Cr [87]	The same	9–18	–	1 W, ~0.3Ti, ~0.3Y_2O_3	~300; l ~12 × 10^3	2–5
SUS316L [57]	Austenitic	17	11	2.5Mo, 1.8Mn, 1TiC	100–300	–
UFG316SS [15, 58]	The same	16.6	10.6	2.25Mo, 1.1Mn, 0.7Si	40	–

complex enough, as shown in Fig. 3.11 for the Japanese F95 steel fabrication [86]. Figure 3.12 illustrates how a steel fibrous structure arises after the combination of the HE, CR and annealing. Such elongated (but shorter and narrower) grains are also characteristic for ferritic steels after HE/R treatment (14YWT, MA957, etc.), but the grains in the austenitic steels (HPT treatment) have roughly an equiaxed structure. As mentioned above, the resource of reactor tests for ferritic and austenitic nanosteels has not been rich yet. A detailed analysis of technology and properties of radiation-resistant ferritic nanosteels is presented in survey [88].

The above listed non-nickel steels are used as cladding for fuel pin in sodium fast reactors, and now they are studied as promising candidates to be used in the International Thermonuclear Experimental Reactor (ITER) and DEMO types, where the operation conditions correspond to the extreme regimes described in Zinkle's and Snead's survey [89]. Figure 3.13 demonstrates a general view of a projected fusion reactor, where at temperature of $\sim 10^8$ °C the thermonuclear synthesis of hydrogen isotopes proceeds by the mechanism:

$$^2D + {}^3T \rightarrow {}^3He \ (3.5 \ \text{MeV}) + n(14.1 \ \text{MeV}) \tag{3.2}$$

with energy release and formation of fast helium ions and neutrons.

It is clear from Fig. 3.13 that the most intensive irradiation zone is so-called blanket (the first wall of tokomak), where the fast neutrons generated in plasma by reaction (3.2) are cooled and their energy transforms to heat, which must be released (and used!) by an abstraction system. Herein, the thermal loads are of ~ 10 MW/m^2

Fig. 3.11 Cladding tube
manufacturing for F95 steel
(adapted from [86])

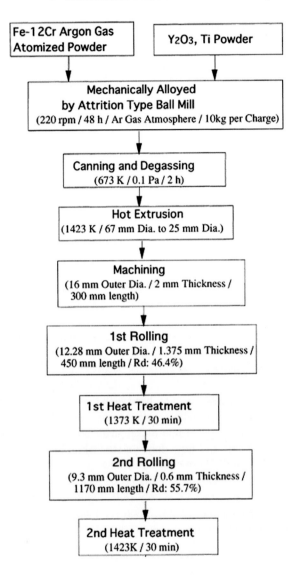

and the irradiation doses can achieve 150–200 dpa, and thus, the most convenient substance as a blanket plasma-facing material is tungsten as the most high-melting and weakly activated metal with high thermal conductivity and strength characteristics. Moreover, not only the tungsten sublimation ability and abrasive wear parameters are very low, but also the tritium isotopes absorption is low. The investigation of W samples under an intensive pulse attack of the high-energy plasma jet (with energy flow densities up to 1 MJ/m^2) has demonstrated that degradation of samples is accompanied by both the surface evaporation/fusion and the destruction of near-surface layers in depth ranges to 150–250 μm [90].

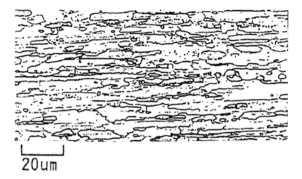

Fig. 3.12 Optical microstructure of cladding tube for F95 steel after second heat treatment (adapted from [86])

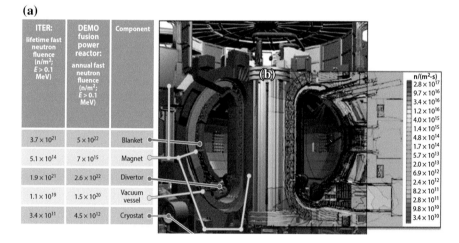

Fig. 3.13 Overview cross section of the ITER tokomak (toroidal camera with magnetic coils): **a** calculated fast neutron fluences for different key components in ITER and DEMO reactors; **b** highlights the position-dependent total neutron fluxes (adapted from [89]). The ITER reactor is now being built in Cadarache (France) with the participation of EU, Japan, USA, Russia, China, India and South Korea

Also, it should be noted that from the standpoint of using tungsten under specific operational conditions of the nuclear units, its two main shortages should be taken into account: brittleness and high value of T_{BDT}. These factors negatively influence the radiation resistance. In order to depress their role, prolonged and diversified investigations were and are carried out mainly on the basis of the nanostructure approaches. The tungsten-based materials were modified by introducing some disperse strengthening additives (such as TiC, Y_2O_3, and La_2O_3); by using various combinations of treating technological methods, such as MA, HIP, the pressure treatment in a superplasticity state, and the formation of different laminated structured composites reinforced by W wires, etc. (e.g., [14, 91–94]). In laboratory conditions, GS was decreased to 50–200 nm, which allowed even examination of plasticity at room temperature and superplasticity at $T = 1650$ °C [92]. The neutron irradiation studies ($E > 0.1$ MeV, $T = 800$–1073 K, 0.15–0.47 dpa) of some blanket plasma-facing materials for several W alloys (with TiC additives or doped with La and K) have shown that their radiation resistance in all cases was higher than that for commonly used tungsten, but the minimum radiation hardening and the low number of radiation defects were observed namely in the UFG W–0.3TiC [93] material, which confirms the earlier obtained results presented in Tables 3.1 and 3.3 [14].

Then, it is presumed that W can be used in a diverter as a supplementary toroidal nuclear mounting part destined to remove the charged particles and blanket wear products. As an alternative promising material for diverters and blankets, the CVI SiC/SiC composites are considered because they stand out as having high radiation and thermal stability (see Fig. 3.7) [63]. Some problems concerning the transmutation processes under the materials irradiation at the nuclear fusion reactor environment are considered in [95].

Various SiC-based materials are also used in other fields of nuclear technologies, for example, as so-called TRISO-coated fuel UO_2 microparticles in superhigh-temperature gas-cooled reactors. Such 30–40 μm thick coatings are applied through CVD method in fluidized-bed layers and act a diffusion barrier for solid and gaseous fission products. It was shown that the minimal porosity (~ 0.04 %), related maximal hardness (~ 50 GPa), and elastic modulus (~ 340 GPa) of these coatings correspond to the deposition temperature about 1450 °C [96]. The presented mechanical properties are close to those for nanocrystalline SiC films [25, 26]. Furthermore, it is known that tubes from impregnated SiC/SiC composites (with a CVI SiC matrix) are investigated as materials promising for replacement of traditional zirconium shells of the rod-type fuel elements in the traditional thermal reactors.

It should be noted that some nanocomposites containing W, ^{10}B, Gd, Hf, Cd, TiB_2, and B_4C (i.e., highly absorptive components with a great neutron trapping cross-section) are investigated in order to create new materials for absorbing and regulative rods in the reactor control systems, as well as for radiation shielding against a neutron and γ-irradiation (e.g., [97–99]). It has been recently shown that some nanostructured luminophors (on the basis of $CaSiO_4$, Al_2O_3, and many other compounds) hold great promises for application in great dose β-radiation dosimetry [100, 101].

The above presented data clearly demonstrate the importance and possible applications of various NMs in the field under investigation as well as the necessity of the further investigations of their radiation stability. Besides, the radiation resistance data are the subject of much current interest in connection with problems of radiation shielding design to protect both the personal and element base of aerospace systems, etc. In general, the experimental and theoretical studies of these materials radiation stability remain very actual and the accumulation of information about the NM irradiation occurs very intensively. Some of the recent most interesting results are presented below:

1. The unexpected fast grain growth in nanocrystalline ceria under 3 MeV Au ions [102];
2. Quantitative comparison of sink efficiency of Cu–Nb, Cu–V and Cu–Ni heterointerfaces for radiation point defects [103];
3. High radiation resistance of UFG 304L stainless steel ($L \sim 100$ nm), produced by ECAP technique, under 3.5 MeV Fe ion irradiation at 500 °C up to dose of 80 dpa [104];
4. Radiation tolerance of nanocrystalline Yttria stabilized Zirconia (10YSZ) increased with decreasing grain size from 220 to 25 nm (irradiation with 400 keV Kr ions up to 129 dpa) [105].
5. High irradiation resistance of nanolaminated MAX phases (Ti_3SiC_2, Ti_3AlC_2, Ti_2AlN, and so on) [106,107].

References

1. Golubov SI, Barashev AV, Stoller RE (2011) Radiation damage theory. In: Konigs R (ed) Encyclopedia of comprehensive nuclear materials. Elsevier, Amsterdam, Chap. 29
2. Rose M, Balough AG, Hahn H (1997) Instability of irradiation induced defects in nanostructured materials. Nucl Instr Meth Phys Res B 127–128:119–122
3. Misra A, Demkowicz MJ, Zhang X et al (2007) The radiation damage tolerance of ultrahigh strength nanolayered composites. JOM 52:62–65
4. Wurster S, Pippan R (2009) Nanostructured metals under irradiation. Scr Mater 60:1083–1087
5. Andrievskii RA (2010) Effect of irradiation on properties of nanomaterials. Phys Met Metallogr 110:229–240
6. Demkowicz MJ, Bellon P, Wirth BD (2010) Atomic-scale design of radiation-tolerant nanocomposites. MRS Bull 35:992–998
7. Andrievski RA (2011) Behavior of radiation defects in nanomaterials. Rev Adv Mater Sci 29:54–67
8. Perry AJ (1998) Microstructural changes in ion implanted titanium nitride. Mater Sci Eng, A 253:310–318
9. Rose M, Gorzawski G, Miche G et al (1995) Phase stability of nanostructured materials under heavy ion irradiation. Nanostruct Mater 6:731–734
10. Nita R, Schaeublin R, Victoria M (2004) Impact of radiation on the microstructure of nanocrystalline materials. J Nucl Mater 329–333:953–957
11. Nita R, Schaeublin R, Victoria M et al (2005) Effect of radiation on the microstructure and mechanical properties of nanostructured materials. Phil Mag 85:723–735
12. Shen TD, Feng Sh, Tang M et al (2007) Enhanced radiation tolerance in nanocrystalline Mg_2GaO_4. Appl Phys Lett 90:263115 (1–3)
13. Kilmametov AR, Gunderov DV, Valiev RZ et al (2008) Enhanced ion irradiation resistance of bulk nananocrystalline TiNi alloy. Scr Mater 59:1027–1030
14. Kurushita H, Kobayashi S, Nakai K et al (2008) Development of ultra-fine grained W–(0.2–0.8)wt% TiC and its superior resistance to neutron and 3 MeV He-ion irradiation. J Nucl Mater 377:34–40

15. Ettienne A, Radiguet B, Cunningham NJ et al (2011) Comparison of radiation-induced segregation in ultrafine-grained and conventional 316 austenitic stainless steels. Ultramicroscopy 111:659–663
16. Yu KY, Liu Y, Sun C et al (2012) Radiation damage in helium ion irradiated nanocrystalline Fe. J Nucl Mater 425:140–146
17. Parish CM, White RM, LeBeau JM et al (2014) Response of nanostructured ferritic alloys to high-dose heavy ion irradiation. J Nucl Mater 445:251–260
18. Fu EG, Misra A, Wang H et al (2010) Interface enabled defects reduction in helium ion irradiated Cu/V nanolayers. J Nucl Mater 407:178–188
19. Misra A, Thilly L (2010) Structural metals at extremes. MRS Bull 35:965–972
20. Gao Y, Yang T, Xue J et al (2011) Radiation tolerance of Cu/W multilayered nanocomposites. J Nucl Mater 413:11–15
21. Wang H, Gao Y, Fu E at al (2014) Irradiation effects on multilayered W/ZrO$_2$ film under 4 MeV Au ions. J Nucl Mater 455:86–90
22. Milosavljević M, Milinović V, Peruško D et al (2011) Stability of nano-scaled Ta/Ti multilayers upon argon irradiation. Nucl Instr Meth Phys Res B 269:2090–2097
23. Hong M, Ren F, Zhang H et al (2012) Enhanced radiation tolerance in nitride multilayered nanofilms with small period-thicknesses. Appl Phys Lett 101:153117 (1–5)
24. Yu KY, Liu Y, Fu EG et al (2013) Comparisons of radiation damage in He ion and proton irradiated immiscible Ag/Ni nanolayers. J Nucl Mater 440:310–318
25. Andrievski RA (2009) Synthesis, structure and properties of nanosized silicon carbide. Rev Adv Mater Sci 22:1–20
26. Wu R, Zhou K, Yue ChY et al (2015) Recent progress in synthesis, properties and potential applications of SiC nanomaterials. Progr Mater Sci 72:1–60
27. Leconte Y, Monnet I, Levalois M et al (2007) Comparison study of structural damage under irradiation in SiC nanostructured and conventional ceramics. Mater Res Soc Symp Proc 981: JJ07 (1–6). (MRS, Warrendale)
28. Jiang W, Wang H, Kim I et al (2009) Response of nanocrysralline 3C silicon carbide to heavy-ion irradiation. Phys Rev B 80:161301 (R) (1–4)
29. Gosset D, Audren A, Leconte Y et al (2012) Structural irradiation damage and recovery in nanometric silicon carbide. Progr Nucl Energy 57:52–56
30. Zhang Y, Ishimaru M, Varga T et al (2012) Nanoscale engineering of radiation tolerant silicon carbide. Phys Chem Chem Phys 14:13429–13436
31. Ishimaru M, Zhang Y, Shannon S et al (2013) Origin of radiation tolerance in 3C-SiC with nanolayered planar defects. Appl Phys Lett 103:033104 (1–4)
32. Meldrum A, Boatner LA, Ewing RC (2002) Nanocrystalline zirconia can be amorphized by ion irradiation. Phys Rev Lett 88:025503 (1–3)
33. Sickafus KE, Matzke H, Hartman T et al (1999) Radiation damage effects in zirconia. J Nucl Mater 274:66–77
34. Johannessen B, Kluth P, Liewellyn DJ et al (2007) Amorphization of embedded Cu nanocrystals by ion irradiation. Appl Phys Lett 90:073119 (1–3)
35. Johannessen B, Kluth P, Liewellyn DJ et al (2007) Ion-irradiation-induced amorphization of Cu nanoparticles embwddwd in SiO$_2$. Phys Rev B 76:184203 (1–11)
36. Kluth P, Johannessen B, Foran GJ et al (2006) Disorder and cluster formation during ion irradiation of Au nanoparticles in SiO$_2$. Phys Rev B 74:014202 (1–8)
37. Ridgway MC, Azevedo GM, Elliman RG et al (2005) Ion-irradiation-induced preferential amorphization of Ge nanocrystals in silica. Phys Rev B 71:094107 (1–6)
38. Djurabekova F, Backman M, Pakarinen OH et al (2009) Amorphization of Ge nanocrystals embedded in amorphous silica under ion irradiation. Instr Meth Phys Res B 267:1235–1238
39. Sprouster DJ, Giulian R., Araujo LL et al (2010) Ion irradiation induced amorphization of cobalt nanoparticles. Phys Rev B 81:1554 (1–8)
40. Krasheninnikov AV, Nordlund K (2010) Ion and electron irradiation-induced effects in nanostructured materials. J Appl Phys 107:071301 (1–70)

41. Chukalkin YuG (2013) Amorphization of oxides by irradiation of fast neutrons. Phys Solid State 55:1601–1604
42. Kaomi D, Motta AT, Birtcher RC (2008) A thermal spike model of grain growth under irradiation. J Appl Phys 104:073525 (1–13)
43. Radiguet B, Etienne P, Pareige P et al (2008) Irradiation behavior of nanostructured 316austenitic stainless steel. J Mater Sci 43:7338–7343
44. Lian J, Zhang J, Namavar F et al (2009) Ion beam-induced amorphous-to-tetragonal phase transformation and grain growth of nanocrystalline zirconia. Nanotechnology 20:245303 (1–7)
45. Marquis EA, Hu R, Rousseau T (2011) A systematic approach for the study of radiation-induced segregation/depletion at grain boundaries in steels. J Nucl Mater 413:1–4
46. Certain A, Kuchibhatla S, Shutthanandan V et al (2013) Radiation stability of nanoclusters in nanostructured oxide dispersion strengthened (ODS) steels. J Nucl Mater 434:311–321
47. Vo NQ, Chee SW, Schwen D et al (2010) Microstructural stability of nanostructured Cu alloys during high-temperature irradiation. Scr Mater 63:929–932
48. Tai K, Averback RS, Bellon Pinko VI et al (2009) Radiation-induced reduction in the void swelling. J Nucl Mater 385:228–230
49. Yang T, Huang X, Wang C et al (2012) Enhanced structural stability of nanoporous zirconia under irradiation of He. J Nucl Mater 427:225–232
50. Bringa EM, Monk JD, Caro A et al (2012) Are nanoporous materials radiation resistant? Nano Lett 12:3351–3355
51. Yu KY, Bufford D, Chen Y et al (2013) Basic criteria for formation of growth twins in high stacking fault energy metals. Appl Phys Lett 103:181903 (1–5)
52. Yu KY, Bufford D, Sun C et al (2013) Removal of stacking-fault tetrahedra by twin boundaries in nanotwinned metals. Nat Commun 4:1377–1384
53. Yu KY, Bufford D, Khatkhatay F et al (2013) In situ studies of irradiation-induced twin boundary migration in nanotwinned Ag. Scr Mater 69:385–388
54. Han W, Fu EG, Demkowicz MJ et al (2013) Irradiation damage of single crystal, coarse-grained, and nanograined copper under helium bombardment at 450 °C. J Mater Res 28:2763–2769
55. Wang H, Gao Y, Fu E et al (2013) Effect of high fluence Au ion irradiation on nanocrystalline tungsten film. J Nucl Mater 442:189–194
56. Yamashita S, Akasaka N, Ohnuki S (2004) Nano-oxide particle stability of 9–12Cr grain morphology modified ODS steels under neutron irradiation. J Nucl Mater 329–333:377–381
57. Matsuoka H, Yamasaki T, Zheng YJ et al (2007) Microstructure and mechanical properties of neutron-irradiated ultra-fine-grained SUS316L stainless steels and electrodeposited nanocrystalline Ni and Ni–W alloys. Mater Sci Eng, A 449–451:790–793
58. Pareige P, Etienne A, Radiguet B (2009) Experimental atamic scale investigation of irradiation effects in CW 316SS and UFG-CW 316SS. J Nucl Mater 389:259–264
59. McClintock DA, Hoelzer DT, Sokolov MA et al (2009) Mechanical properties of neutron irradiated nanostructured ferritic alloy 14YWT. J Nucl Mater 386–388:307–311
60. McClintock DA, Sokolov MA, Hoelzer DT et al (2009) Mechanical properties of irradiated ODS-EUROFER and nanocluster strengthened 14YWT. J Nucl Mater 392:353–359
61. Nanstad RK, McClintock DA, Hoelzer DT et al (2009) High temperature irradiation effects in selected Generation IV structural alloys. J Nucl Mater 392:331–340
62. Miller MK, Hoelzer DT (2011) Effect of neutron irradiation on nanoclusters in MA957 ferritic alloys. J Nucl Mater 418:307–310
63. Katoh Y, Nozawa T, Snead LL et al (2011) Stability of SiC and its composites at high neutron fluence. J Nucl Mater 417:400–405
64. Shamardin VK, Goncharenko YD, Bulanova TM et al (2012) Effect of neutron irradiation on microstructure and properties of austenitic AISI 321 steel, subjecting to equal-channel angular pressing. Rev Adv Mater Sci 31:167–173
65. Alsabbagh A, Valiev RZ, Murty KL (2013) Influence of grain size on radiation effects in a low carbon steel. J Nucl Mater 443:302–310

66. Koyanagi T, Shimoda K, Kondo S et al (2014) Irradiation creep of nanopowder sintered silicon carbide at low neutron fluences. J Nucl Mater 455:73–80
67. Ovid'ko IA, Sheinerman AG (2005) Irradiation-induced amorphization processes in nanocrystalline solids. Appl Phys A 81:1083–1088
68. Shen TD (2008) Radiation tolerance in a nanostructure: is a smaller better? Nucl Instr Meth Phys Res B 266:921–925
69. Oksengendler BI, Turaeva NN, Maximov SE et al (2010) Peculiarities of radiation-induced defect formation in nanocrystals imbedded in a solid matrix. J Exp Theor Phys 111:415–420
70. Samaras M., Derlet PM, Van Swygenhoven H et al (2002) Computer simulation of displacement cascades in nano crystalline Ni. Phys Rev Lett 88:125505 (1–4)
71. Voegeli W, Albe K, Hahn H (2003) Simulation of grain growth in nanocrystalline nickel induced by ion irradiation. Nucl Instr Meth Phys Res B 202:230–235
72. Samaras M, Derlet PM, Van Swygenhoven H et al (2003) SIA activity during irradiation of nanocrystalline Ni. J Nucl Mater 323:213–219
73. Mayr SG, Averback RS (2003) Evolution of morphology in nanocrystalline thin films during ion irradiation. Phys Rev B 68:075419 (1–9)
74. Samaras M, Derlet PM, Van Swygenhoven H et al (2006) Atomic scale modeling of the primary damage state of irradiated FCC and BCC nanocrystalline metals. J Nucl Mater 351:47–55
75. Millet PC, Aidhy DS, Desai T et al (2009) Grain-boundary source/sink behavior for point defect: an atomistic simulation study. Int J Mater Res 100:550–555
76. Beyerlein IJ, Caro A, Demkowicz et al (2013) Radiation damage tolerant nanomaterials. Mater Today 16:443–449
77. Morishita K, Watanabe Y, Kohyama A et al (2009) Nucleation and growth of vacancy-clusters in β-SiC during irradiation. J Nucl Mater 386–388:30–32
78. Swaminathan N, Kamenski PJ, Morgan D et al (2010) Effects of grain size and grain boundaries on defect production in nanocrystalline 3C-SiC. Acta Mater 58:2843–2853
79. Jiang H, Jiang C, Morgan D et al (2014) Accelerated atomistic simulation study on the stability and mobility of carbon tri-interstitial cluster in cubic SiC. Comp Mater Sci 89:182–188
80. Psakhie SG, Zolnikov KP, Kryzhevich DS et al (2009) Evolution of atomic collision cascade in vanadium crystal with internal structure. Crystal Rep 54:1002–1011
81. Bai X-M, Voter AF, Hoagland RG et al (2010) Efficient annealing of radiation damage near grain boundaries via interstitial emission. Science 327:1631–1634
82. Yang Y, Huang H, Zinkle SJ (2010) Anomaly in dependence of radiation-induced vacancy accumulation on grain size. J Nucl Mater 405:261–265
83. Bai X-M, Uberuaga BP (2012) Multi-timescale investigation of radiation damage near TiO$_2$ rutile grain boundaries. Phil Mag 92:1469–1498
84. Chimi Y, Iwase A, Ishikawa N et al (2001) Accumulation and recovery of defects in ion-irradiated nanocrystalline gold. J Nucl Mater 297:355–357
85. Zinkle SJ, Was GS (2013) Materials challenges in nuclear energy. Acta Mater 61:735–758
86. Ukai S, Mizuta S, Yoshitake T et al (2000) Tube manufacturing and characterization of oxide dispersion strengthened ferritic steels. J Nucl Mater 283–287:702–706
87. Dubuisson P, Carlan Y, Garat V et al (2012) ODS ferritic/martensitic alloys for sodium fast reactor fuel pin cladding. J Nucl Mater 428:6–12
88. Odette GR, Alinger MJ, Wirth BD (2008) Recent developments in irradiation-resistant steels. Annu Rev Mater Res 38:471–503
89. Zinkle SJ, Snead LL (2014) Designing radiation resistance in materials for fusion energy. Annu Rev Mater Res 44:241–287
90. Voronin AV, Sud'enkov YV, Semenov BN et al (2014) Degradation of tungsten under the action of a plasma jet. Tech Phys 59:981–988
91. Rieth M, Dudarev SL, Gonzales de Vicente SM et al (2013) A brief summary of the progress on the EFDA tungsten materials program. J Nucl Mater 442:173–180

92. Wurster S, Baluc N, Battabyal M et al (2013) Recent progress in R&D on tungsten alloys for divertor structural and plasma facing materials. J Nucl Mater 442:181–189

93. Efe M, El-Atwani O, Guo Y et al (2014) Microstructure refinement of tungsten by surface deformation for irradiation damage resistance. Scr Mater 70:31–34

94. Fukuda M, Hasegawa A, Tanno TS et al (2013) Property change of advanced tungsten alloys due to neutron irradiation. J Nucl Mater 442:273–276

95. Sawan ME, Katoh Y, Snead LL (2013) Transmutation of silicon carbide in fusion nuclear environment. J Nucl Mater 442:370–375

96. Kim W-J, Park JN, Cho MS et al (2009) Effect of coating temperature on properties of the SiC layer in TRISO-coated particles. J Nucl Mater 392:213–218

97. Buyuk B, Tugrul B, Akarsu AC et al (2011) Investigation on the effects of TiB_2 particle size on radiation shielding properties of TiB_2 reinforced BN–SiC composites. In: Pogrebnjak AD (ed), Nanomaterials: Applications and Properties (NAP-2011, Alushta), Ukrainian Sumy University, Sumy, vol 2, part II, pp 421–428

98. Schrempp-Koops L (2013) Size efficiency of neutron shielding in nanocomposites—a full-range analysis. Int J Nanosci 12:1350015 (1–8)

99. Kim J, Seo D, Lee BC et al (2014) Nano-W dispersed gamma radiation shielding materials. Adv Eng Mater 16:1083–1089

100. Salsah N (2011) Nanocrystalline materials for the dosimetry of heavy charged particles: a review. Rad Phys Chem 80:1–10

101. Kortov VS, Nikiforov SV, Moiseikin EVAG et al (2013) Luminescent and dosimetric properties of nanostructured ceramics based on aluminium oxide. Phys Sol State 55:2088–2093

102. Aidhy DS, Zhang Y, Weber WJ (2014) A fast grain-growth mechanism revealed in nanocrystalline ceramic oxides. Scr Mater 83:9–12

103. Mao Sh, Shu SH, Zhou J et al (2015) Quantitative comparison of sink efficiency of Cu–Nb, Cu–V and Cu–Ni interfaces for point defects. Acta Mater 82:328–335

104. Sun C, Zheng S, Wei CC et al (2015) Superior radiation-resistant nanoengineered austenitic 304L stainless steel for applications in extreme radiation environments. Sci Rep 5:7801 (1–6)

105. Dey S, Drazin JW, Wang Y et al (2015) Radiation tolerance of nanocrystalline ceramics: insight from Yttria stabilized Zirconia. Sci Rep 5:7746 (1–9)

106. Tallman DJ, Hoffman EN, Caspi EN et al (2015) Effect of neutron irradiation on select MAX phases. Acta Mater 85:132–141

107. Huang Q, Liu R, Lei G et al (2015) Irradiation resistance of MAX phases Ti_3SiC_2 and Ti_3AlC_2: characterization and comparison. J Nucl Mater 465:640–647

Chapter 4
Mechanical Actions Effect upon Nanomaterials

Abstract The main SPD methods for microstructure refinement and their influence on the on the nanostructure formation are briefly considered. The data on cyclic and other loadings (including a combined action) on the structure, phase transitions and properties of NMs are also analyzed. Some theoretical approaches and the MD simulation method possibilities are described. The importance of nanotwinned gradient surface structures in increasing mechanical properties is underlined. Examples of the mechanical action practical applications are given. A special attention is paid to the poorly studied problems in the field.

4.1 General Considerations

The influence of mechanical actions on the nanostructures is multi-versioned, because NMs have to work in various operational situations under static and dynamic loads in a wide range or even under combination of different extreme conditions. From the other side, the SPD methods for grain refinement, such as ECAP, HPT, HE/R, FMRR, etc., are also widely used in the NMs processing. In this connection, it is possible to add to this list some other technologies, such as an accumulative roll bonding (ARB), screw rolling, cold rolling (CR), surface mechanical grinding treatment (SMGT), and surface rolling treatment (SRT). These methods are schematically depicted in Fig. 4.1 [1–3].

The so-called effective deformation ε_{eff} for some listed SPD methods can be estimated using the expressions [1]:

$$\varepsilon_{\text{eff}} = n2/\sqrt{3} \, \cos\varphi \, (\text{for ECAP}), \tag{4.1}$$

$$\varepsilon_{\text{eff}} = n2/\sqrt{3}\pi r/t \, (\text{for HPT}), \tag{4.2}$$

$$\varepsilon_{\text{eff}} = n2/\sqrt{3} \, \ln t_0/t \, (\text{for ARB}), \tag{4.3}$$

© Springer International Publishing Switzerland 2016
R.A. Andrievski and A.V. Khatchoyan, *Nanomaterials in Extreme Environments*,
Springer Series in Materials Science 230, DOI 10.1007/978-3-319-25331-2_4

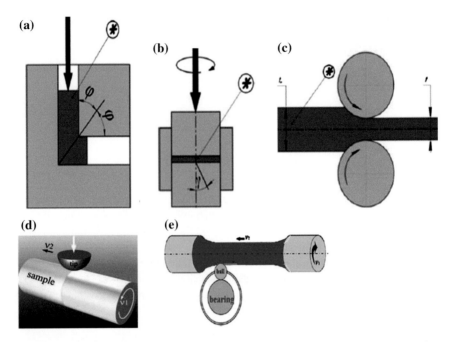

Fig. 4.1 Schematic illustration of some SPD techniques: **a** ECAP. **b** HPT. **c** ARB (* is a designation for initial sample; adapted from [1]). **d** SMGT (adapted from [2]. **e** SRT (adapted from [3])

where n is the number of passes or revolutions, r is the distance from the axis, t and t_o are the running and initial thickness of sample. By convention for SPD treatment, the most characteristic are regimes providing the values $\varepsilon_{eff} \geq 1$ (absolute deformation $\varepsilon \geq 100$ %).

The traditional technology schemes, such as ECAP, HPT, and ARB, are directed to obtaining more or less uniform nano- and UFG structures all over the sample cross-section, whereas the advanced SMGT and SRT methods were specially developed in order to create a gradient surface structure described in details in [2, 3]. Figure 4.2 shows the grain/subgrain size as a function of the distance from the surface of Ti bar (16 mm in diameter, the average grain size about ~ 30 µm) after SRT (adapted from [3]).

As illustrated in Fig. 4.2, the nanostructure with GS below 100 nm is observed after SRT only at distances from surface don't exceeding of ~ 70 µm (region I), where deformations occurs predominantly at the sacrifice of twinning, but with GS growth (in regions II and III) the low-angle TBs transform into high-angle ones and the deformation acquires a dislocation nature. For these processes, the total extension of deformation zone is about 10^3 µm (i.e., about 12 % of the treated samples radius). As an example, the main mechanical properties for Ti samples at the initial state and after various treatment procedures (SRT, ECAP, ARB, and others) are presented in Table 4.1.

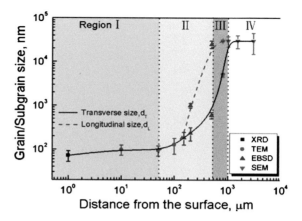

Fig. 4.2 The grain/subgrain size distribution along the distance from the surface according to XRD, TEM, EBSD, and SEM data

Table 4.1 Mechanical properties of Ti before and after different technology treatments

Treatment	Sample	Grain size (nm)	σ_{UTS} (MPa)	σ_Y (MPa)	δ (%)
SRT [3]	Initial	$3 \cdot 10^4$	451	380	24
	Final	See Fig. 4.2	495	437	21
ECAP [4]	Initial	$3 \cdot 10^4$	700	530	25
	Final	~200	1240–1250	1000–1250	11–13
ARB at low temperatures [5]	Initial	n/d	420	300	26
	Final	35–70	700–945	665–860	6–11
Screw and sizing length-wise rolling [6]	Initial	n/d	490	376	29
	Final	180–190	850–905	670–722	10–12.5

σ_{UTS} is an ultimate tensile stress, σ_Y is an yield stress, and δ is an elongation to fracture

As is evident from these data, the general growth of the NMs mechanical properties due to the treatment is sufficiently below that for SRT method (in comparison with other SPD methods), and this fact can be easily explained by small extension of the deformation zone. From the other side, the noticeable growth of surface microhardness (by nearly 3 times) and high ductility of the materials can be very useful for many applications.

Also, it should be noted that additional loading of a rotating sample with a moving hard tip (see Fig. 4.1e) provides the generation of surface layers with higher properties due to the strongly textured nanolaminated structure (NLS) formation with an average lamellar thickness of 20 nm, as it was shown for Ni samples [2]. The deformation zone extension in 10 mm diameter bars was about 1 mm (i.e., 20 % of the sample radius); herein, the surface microhardness was equal to 6.5 GPa, which is sufficiently higher than the H_V values for the samples produced by other SPD methods not to mention the CG nickel (Fig. 4.3).

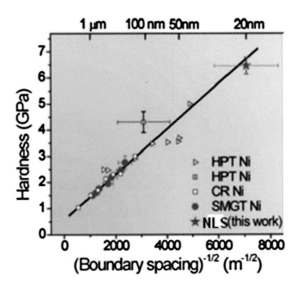

Fig. 4.3 Hall-Petch plot for the HPT, SMGT, NLS, and CR nickel specimens (adapted from [2])

Also, it is important that the samples with NLS have also demonstrated the higher thermal stability. As distinct from the many data for nanocrystalline nickel, for which the dependence of the hardness on GS exhibits a non-monotonic character (e.g., [8]), the results for NLS Ni are in good agreement with the well-known classic Hall-Petch relationship:

$$H = H_o + K_{HP}L^{-1/2}, \tag{4.4}$$

where K_{HP} is the Hall-Petch coefficient and H_o is a constant associated with the single crystal hardness. Analogous data concerning the consistency of the results with (4.4) were obtained for the NLS interstitial-free iron samples treated by SMGT method [7]. The surface microhardness (at the mean lamella thickness about 20 nm) was of 5.3 ± 0.6 GPa.

It is commonly supposed that the experimental results inclination from relation (4.4) is connected with a limited number of the lattice dislocations, because their formation into nanostructure little grains is impossible [8–10], and as a result, instead of the usual deformation due to dislocation shifts, there some other mechanisms arise, such as intergrain sliding or diffusion creep. It is interesting that here the strength decreases with decreasing GS, which corresponds to the so called inversed Hall-Petch dependence (sometimes called even an "anti-HP" dependence), because the presence of the internal nanolaminates (and correspondingly TBs) promotes strength growth. Figure 4.4 shows the nanoscale TB strengthening mechanism-based dislocation-TB interaction.

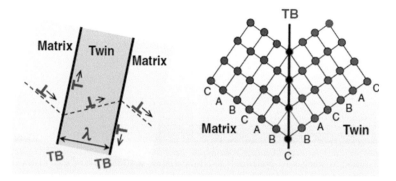

Fig. 4.4 Schematic illustration of the nanoscale TB strengthening mechanism (adapted from [11])

It is assumed that the interaction between full dislocations and twins leads to the formation of Shockley partially dislocations (with the Burger vector $b = 1/6[1\bar{2}1]$), easily moving on the coherent boundaries, and no non-glissile Frank dislocations ($b = 1/3[111]$) under dislocation dissociation reactions, such as $1/2[101] \rightarrow 1/6[1\bar{2}1] + 1/3[111]$. It is believed that glissile Shockley partial dislocations provide plasticity and non-glissile Frank dislocations are responsible for the strength. The formation of Shockley partial dislocation is confirmed by the HRTEM results.

A theoretical model of the plastic flow through widening of nanoscale twins in NMs was developed by Morozov et al. [12], where the authors supposed disclinations formation in the contact points between twins and GBs. The comparison of the calculated and experimental [11] data for the samples with lamella width λ in the 4–14 nm interval demonstrated that they are in a close agreement with each other.

The generation of nanotwins and low-angle structures can be connected with both the samples production processes (so-called growth twins) and peculiarities of deformation itself (deformation twins) or recrystallization annealing (anneal twins). The use of ARB technology (Fig. 4.1c) proved to be very efficient for creation of a regular array of atomic-scale facets (i.e., the zig-zag morphology) in Cu–Nb nanolayered materials [13]. The two types of ordered interfaces after very high strains decreasing the initial thickness by 5–6 orders of magnitude (from 2 mm to 20 nm!) are shown in Fig. 4.5. The experiments also demonstrated that the interfaces in Cu–Nb nanolaminates shown in Fig. 4.5 are stable with respect to continued extreme straining as well as high-temperature and irradiation effects (see Chaps. 2 and 3).

It is worthy to mark a role of nanotwinned structures in the growth of some important characteristics, such as the hardness H_V, fracture toughness K_{IC} and onset oxidation temperature T_{oxid} for some typical superhard materials, namely diamond and cubic BN (Table 4.2).

The samples of c-BN and diamond with the nanotwinned structure were produced by a high-pressure/high-temperature method ($P = 10$–25 GPa and $T = 1800$–2000 °C) using initial particles with onion-like structures [14, 15]. The data of Table 4.2 apparently demonstrate an advantage of the nanotwinned structures, but

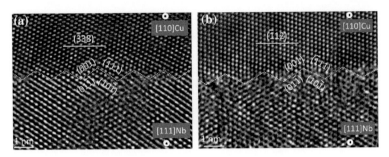

Fig. 4.5 HRTEM images of preferred Cu–Nb interfaces: **a** {338}<443>Cu ‖ {112}<110>Nb. **b** {112}<111>Cu ‖ {112}<110>Nb (adapted from [13])

Table 4.2 Properties of nanotwinned/nanocrystalline cubic BN [14] and diamond [15]

Subject	Structure		H_V (GPa)	K_{IC} (MPA m^{½})	T_{oxid} (°C)
	Grain size (nm)	Lamella thickness (nm)			
Nanocrystalline c-BN	~14	–	85	6.8	~1100
Nanotwinned c-BN	–	3.8	~110	12.7	~1300
Nanocrystalline diamond	10–30	–	110–140	5–15	~800
Nanotwinned diamond	–	~5	175–204	9.7–14.8	~1000

the details of possible arising of the material properties, such as K_{IC} and T_{oxid} (especially as applied to brittle NMs), need the further studies and considerations.

4.2   Main Experimental Results

4.2.1   Fatigue

The possibility to markedly increase the strength of the materials at the expense of grain refinement took an active interest to their fatigue characteristics analysis [1, 16, 17]. For example, let us consider the results for the ECAP technique influence on the fatigue strength for Cu and Ti [4, 18] presented in Fig. 4.6, where the values of the CG samples are also included for comparison.

From these data, it is evident that for ordinary fatigue tests with 10^4–10^7 cycles the advantages of Cu and Ti UFG samples preserve, but with N growth it diminishes for copper and even grows for titanium. The characteristics of austenitic chromium-nickel steel after ECAP technique treatment in the same cyclic tests

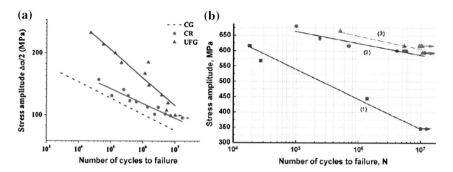

Fig. 4.6 The stress amplitude dependencies on the number of cycles to failure (the so- called Wöhler's *S–N* curves) for samples of Cu (**a**) (adapted from [18]) and Ti (**b**) (adapted from [4]): **a** CG samples ($L \sim 150$ μm); CR ones (several hundred μm in length and dozen μm in width); UFG ones ($L_1 \sim 500$ nm (grains with low-angle boundaries) + $L_2 \sim 2$ μm (grains with high-angle boundaries). **b** (*1*) CG ones ($L \sim 30$ μm); (*2*) those after ECAP ($L \sim 200$ nm); (*3*) those after ECAP + annealing at 350 °C ($L \sim 200$ nm)

range are very similar, but there was fixed the formation of some subgrain structures with the element sizes about 100–250 nm as well as the origination of a large share of deformation twins and lath martensite [19]. A proper interpretation of various peculiarities for the many measured dependences of stress amplitude/cycles number to failure is a very hard problem, because such dependences (so-called *S–N* curves) as applied to different metals and alloys are connected with many effects, such as the deformation temperature, material surface texture and roughness, cracking state, grain growth under given cyclic loads, the presence of low- and high-angle boundaries, admixtures and additions, etc. [17, 20]. In addition, these factors are not usually described in publications correctly and in detail.

The nanostructure evolution under failure tests is thoroughly considered in [18, 19]. Especially for copper, it was marked that for a cyclic torsion loading regime, as distinct from the results of tension↔compression tests (Fig. 4.6a), the *S–N* dependences lower in a parallel way but still demonstrate and preserve some distinctions between materials obtained by different methods (CG, CR, and UFG treatment) throughout the entire range of *N* [18]. The more detailed studies of GS have revealed that the GS markedly increases in the sample periphery, but the structure of the central region remains unchanged and this defines the *S–N* curves characteristics (Fig. 4.6a). It is clear that the fatigue strength depends mainly on the microcracks generation and the peculiarities of development, which, in their turn, are connected with the material stressed state, defects, intergrain boundaries characteristics, etc.

The studies of the fatigue strength of thin layered films and nanowires have shown that the main parameter (cycles number to failure) grows with reducing the Cu films thickness, whereas the Cu–Si nanowires preserve their high strength after cyclic loading tests [21, 22]. The studies of film thickness and GS (in a micrometric

Fig. 4.7 Effect of number of
turns at HPT on fatigue life
and Vickers hardness for Zn–
22 % Al alloy (adapted from
[26])

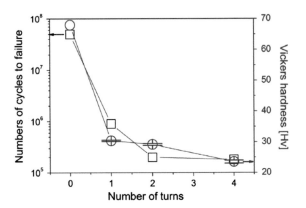

range of 7–170 μm) have shown that the copper fatigue strength grows with
decreasing GS and depends on the film thickness as well [23].

A particular attention was given to the NMs behavior under high-cycle fatigue
tests when the microcracks generation at intergrain boundaries can be enlarged
(e.g., [24]). The fatigue tests (in the range of cycles number $N = 10^8$–2×10^{10}) of
ECAP treated copper samples with GS of 300 nm have demonstrated that their
fatigue strength is twice that of the usual CG copper [25], but from the other side,
the turns number growth during HPT treatment, i.e., the deformation level growth
according to expression (4.2) and the GS decrease, results in decreasing N and H_V
values (Fig. 4.7) [26].

The GS in the eutectoid Zn–22 % Al alloy has the following values: in the initial
state, the GS value is of 1400 nm; after 1 HPT turn, the L value is of 400 nm; after 2
HPT turns, it is of 370 nm; and after 4 HPT turns, −350 nm [26]. It is of interest that
the strength is decreasing with the GS reducing, because it is characteristic for the
superplastic alloys (e.g., see [8]), and in future it must be studied how this picture of
the fatigue strength decrease with the GS reducing (revealed for the Zn–22 % Al
superplastic alloy) may be characteristic for other metals and alloys. In general, it
should be note that it takes long studies to understand the NMs behavior under
high-cycle fatigue.

4.2.2   Phase Transitions

It is well known that under high pressures a crystal lattice can become unstable,
which leads to deformation phase transitions with arising structures characterized
by new symmetry. In the case of NMs, one must especially keep in mind the size
dependence of the phase stability and the appearance of new phase nuclei. For
example, the GS value, below which the martensite transformation under cooling is
impossible for the $Ti_{50}Ni_{25}Cu_{25}$ alloy, equals 15–25 nm (i.e., lower than the critical
size of the martensite phase nucleus), that is, for this alloy at GS below 15–25 nm,

Fig. 4.8 The ω phase fraction plotted versus the average GS for three Ti samples processed by HPT at room and cryogenic temperatures (adapted from [29])

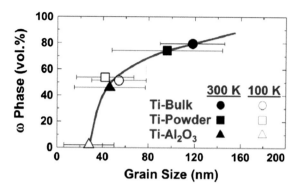

the martensite transformation is retarded simply for the impossibility of this phase nucleus initiation [27].

The size trend for the pressure-induced γ–Fe_2O_3 to α–Fe_2O_3 structural phase transition has been investigated by Alivisatos et al. [28]. The transition pressure was found to increase with decreasing GS: annealed bulk samples transformed at 24 ± 2 GPa, 7 nm crystals transformed at 27 ± 2 GPa, 5 nm ones- at 34 ± 3 GPa, and 3 nm ones- at 37 ± 2 GPa. The authors [28] mention that the transition pressure increase was also revealed in other objects, such as PbS, ZrO_2, Si, TiO_2, etc., and can be connected with both kinetic and thermodynamic reasons as well. In particular, this increase in nanocrystals can be attributed to both the surface energy contribution and features of nucleation points for phase transition.

In order to investigate the phase transitions and changes in the material composition in NMs under mechanical actions, HPT technique is often used. For example, in Fig. 4.8, the dependences of the ω phase volume on GS are presented for the $\alpha \rightarrow \omega$ transitions in Ti for samples subjected to HPT treatment at room and cryogenic temperatures ($P = 6$ GPa, turns number $n = 10$) [29]. It is clear that the ω phase formation markedly decrease with decreasing the initial material GS and HPT treatment temperature. At values $L < 30$ nm and $T = 100$ K, the $\alpha \rightarrow \omega$ transition is practically suppressed completely. Like the above described case of the $Ti_{50}Ni_{25}Cu_{25}$ alloy [27], this can also be connected with an impossibility of the generation of the ω phase nuclei.

The HPT influence on the material properties and content was studied by Straumal et al. [30–32] for the zinc oxides, as well as for the Fe–C and Al–Zn alloys with the GS variation. It was shown that manganese and cobalt solubility in ZnO sufficiently increases with the L values decrease through the HPT action [30], whereas unstable ferrous carbides (such as Fe_5C_2) and residual austenite in nanocrystalline Fe–1.7 wt%C alloys disappears after HPT [31]. In the UFG Al–30 wt%Zn alloy (~ 400 nm) after HPT, the presence of some ~ 2 thick zinc layers at the Al intergrain boundaries were detected by analytical TEM methods (Fig. 4.9) [32]. Such intergrain layers formation can be considered as an example of the so-called partial solid wetting, which does not occur under usual heating followed by annealing, but manifests itself after HPT.

Fig. 4.9 HRTEM image of
Zn layer between Al grains in
UFG Al–30 wt%Zn alloy
after HPT (P = 5 GPa,
T = 300 K, n = 5 turns)
(adapted from [32])

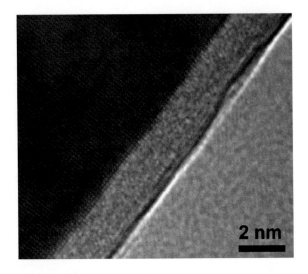

Fig. 4.10 The effect of the
turns at HPT (P = 4 GPa,
T = 300 K) on the crystalline
phase volume fracture for
$Ti_{50}Ni_{25}Cu_{25}$ alloy

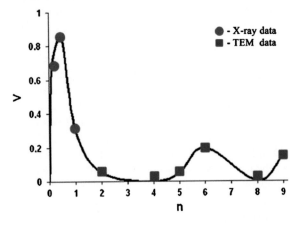

Some examples of deformation-induced amorphization are presented and considered in survey [33]. Thus, an ARB of the multi-layer Cu–Zr composites is
accompanied with an amorphization that can be connected with both a high dislocation density and mutual solubility.

An interesting example of the multi-stage wave-like transformation with a
transition sequence A → N → A → N → A → N (where A and N mean
amorphous and nanocrystalline state, respectively) was observed in the
$Ti_{50}Ni_{25}Cu_{25}$ metallic glass after HPT [34]. In Fig. 4.10, the crystalline phase
volume fracture (V) change is shown as a function of the turns including part of
them such ¼ and ½ (adapted from [34]).

It follows from the Fig. 4.10 data that nanocrystallization of the amorphous
phase begins at very low deformation degree after which a reverse process starts
with wave-form repeats. The authors [34] suggested that such a cyclic character of

the A ↔ N transformation is connected with the dissipation peculiarities of the mechanical energy which was introduced (in other words, "pumped in") into the samples during HPT. Various processes of the phase transformations, defects reorganization and dynamic recrystallization, as well as thermal effects can serve as channels for this energy dissipation mechanism. In the every particular case, the mechanisms depend on numerous factors, such as the temperature, pressure, amorphization thermal barrier, etc. [35], and therefore so far the A ↔ N cyclic transformations remains out of correct predictions.

In several studies, nanosized TiO_2 powders (L = 8–20 nm) during impact compression tests were subjected to high pressure actions (P = 10–45 GPa, T = 500–2500 K) with a load duration of several microseconds, and it was found that the initial material tetragonal phase (anatase) transforms into either the orthorhombic phase (columbite) or the microcrystalline rutile [36]. For comparison with the TiO_2 behavior under high static pressures, it is desirable to show the results obtained by synchrotron XRD and Raman spectroscopy (Fig. 4.11) [37].

As can bee seen in Fig. 4.11a, nanocrystals of GS below of 10 nm undergo pressure-induced amorphization and remain amorphous state (a–TiO_2) upon further compression and decompression. In the 12–50 nm range, the transformation into the monoclinic baddeleyite structure (m–TiO_2) is observed upon compression, which then transforms into the orthorhombic phase (o–TiO_2) on decompression. Coarser crystallites with GS more of 50 nm transform directly to thermodynamically stable o–TiO_2 phase (TiO_2-**II** with α–PbO_2 structure). Two compression-decompression paths of the two samples of study [37] indicated. The bulk TiO_2 samples (Fig. 4.11b) has own pressure-induced features.

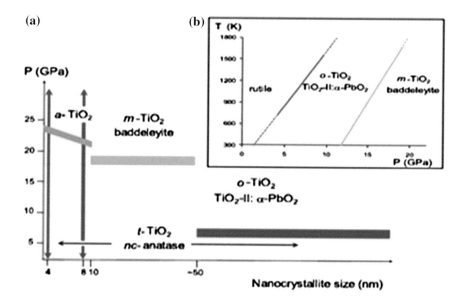

Fig. 4.11 Size-dependent pressure stability of nano TiO_2 (**a**) and known P–T relations for the bulk TiO_2 samples (**b**) (adapted from [37])

4.2.3 Other Examples of Mechanical Actions and Combined Effects

A wide variety of the behavior of NMs (such as Ni, Fe–Cu, CdSe, Al_2O_3, CeO_2, CuO, AlN, etc.) under high static pressures has been generalized by Singh and Kao [38]. It was demonstrated that the volume expansion behavior as a function of pressure up to 50–150 GPa can be described by equations of state using data on the bulk modulus change, and there is a satisfactory agreement between the theory and experimental results, especially in the low pressure range.

Investigations of dynamic recrystallization (i.e., grain growth under various deformations) are very popular. However, most publications are devoted to micromaterials with an initial CG crystal structure. We shall only mention some non-trivial investigations. For example, authors [39] considered the evolution of the high-angle GBs formation during HPT at room temperature and the accompanied dynamic recrystallization in some Fe–Ni alloys. In connection with this, an important observation [40] must be mentioned: a dynamic recrystallization markedly proceeds at room temperature (GS growth from 300 nm to 100 μm) in the crack regions during the rotating–bending fatigue testing processes in UFG copper. These data confirm the before presented results concerning the GS influence on the S–N curves characteristics (Fig. 4.6a) [18].

Two other important peculiarities of the NMs behavior under indentation must especially be marked: a plastic deformation of some brittle materials [8, 41] and their different character at indentation [8, 42]. In Fig. 4.12, the TiN film fracture surfaces near indentation imprints are shown. They have a clear-cut columnar structure, whereas the break as whole has a character of brittle destructions (Fig. 4.12a), but in some places there are pronounced bends of the columns, arising naturally under the volume non-uniform compression (Fig. 4.12b).

The study of the imprint surfaces formed after the nanostructured films indentation using HRSEM methods has demonstrated the two deformation types or mechanisms: a non-uniform one with formation of shift bands and steps (TiB_2; Fig. 4.13a, b) and a uniform one (TiN; Fig. 4.13c, d). The appearance of plasticity

(a)   **(b)**

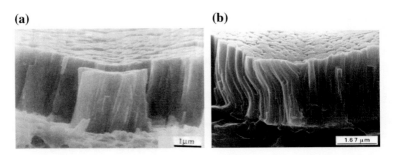

Fig. 4.12 HRSEM images of the TiN film fracture surfaces in the case of brittle (**a**) and residual plastic (**b**) deformation (adapted from [41])

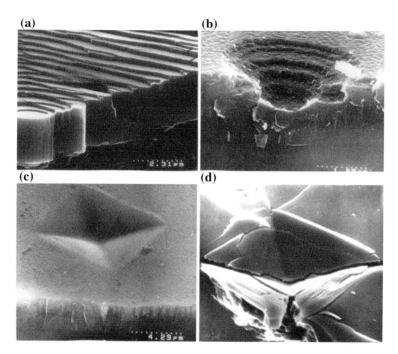

Fig. 4.13 Indentation imprint surfaces in TiB$_2$ (**a** and **b**) and TiN (**c** and **d**) (adapted from [42])

Table 4.3 Effect of ion ^{238}U ($E = 11.4$ MeV) irradiation and high pressure in the diamond anvil cells (P up to 70 GPa) on the ZrO$_2$ structure (adapted from [43, 44])

ZrO$_2$ structure	Initial structure	Fluence (ions/cm^2)	Pressure (GPa)	Final structure
Microcrystalline	Mono	10^{11}–10^{12}	Ambient	Mono
	Mono	10^{13}	Ambient	Tetra
	Ortho-I	$2 \cdot 10^{12}$	10	Cubic
	Ortho-I	$1.5 \cdot 10^{12}$	23	Ortho-II
	Ortho-II	$1.5 \cdot 10^{12}$	38–70	Ortho-II
Nanocrystalline	Mono	10^{13}	Ambient	Tetra

in brittle materials and difference in their deformation nature can be connected with the special peculiarities of the columnar structure of TiN films (Fig. 4.12) and their sliding over the interfaces [8, 41, 42], whereas the TiB$_2$ films are characterized mainly by the absence of the columnar structure and therefore deformation has a more localized and non-homogenous character.

Let us consider some results concerning a combination of the action of several extreme conditions on NMs. The combined action of high pressure and irradiation was studied by the example of ZrO$_2$ and HfO$_2$ in [43, 44]. In Table 4.3, some data are presented for the zirconium dioxide initial and final structures (abbreviations "mono, tetra, cubic, ortho" in the Table 4.3 correspond to the monoclinic, tetrag-onal, cubic, and orthorhombic modifications, respectively).

As it is evident from Table 4.3, there is no difference in the behavior of micro- and nanocrystalline samples. However, the pressure increase in the case of the microcrystalline samples appears causes a change in the final structure, and this was analyzed on the basis of the sizes of the arising lattice distortions owing to the combined action of pressure and irradiation. It was marked that irradiation provokes the phase transition to the high-temperature ortho-II modification which can be annealed and is supposed to be a very radiation-resistant.

A combination of extreme loads was also studied by Fox-Rabinovich et al. [45] using the high-speed cutting treatment of examples. As shown in Fig. 4.14a (adapted from [45]), herein, at the chip/cutting tool interfaces, the temperature can exceed 1000 °C and the pressure achieves 1.5 GPa. The multi-layer coating advantages are obvious from Fig. 4.14b, and they can be connected with a more active formation of a barrier tribo-film, which retards the heat flows propagation and preserves the tool strength, when the TiAlCrSiYN/TiAlCrN multi-layer coating is used. The perspectives of multi-layer nitride coatings composed of alternating polycrystal CrAlN and nanocomposite CrAlSiN layers for the achievement of high strength and toughness were marked in survey [46].

At last, let us consider a very illustrative example of a polyfunctional nanocomposite Cu–Nb wire, used in the solenoids creating pulse super high magnetic fields (>60T) [47, 48]. The demands for such material properties are very high and include the following parameters: the strength not below 1 GPa (to resist to a great Lorenz force arising in magnetic fields); electrical conductivity not below 0.6 the value for pure copper (for the ohmic losses minimization); operational temperature in the 77–673 K range; fatigue durability over 5000 cycles; the radi-ation stability up to He ions flow densities above 10^{15} ions/cm^2. The structure, meeting these tough demands, was created by a complex SPD treatment including repeated HE/R processing, cold drawing, and bundling cycles. Figure 4.15

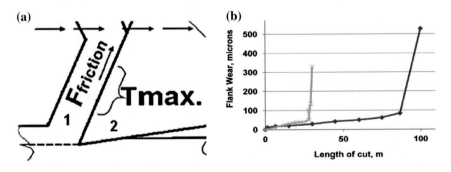

Fig. 4.14 The scheme of a chip(1)/tool(2) interface during cutting (**a**) and tool life of TiAlCrSiYN-based one (*green colored times symbol*)- and multi-layer (*blue colored filled square*) coatings versus cutting speed of 600 m/min (**b**). The total thickness of these coatings was similar (about 2 μm); the GS in one- and multi-layer coatings was below 40 and 20–40 nm, respectively

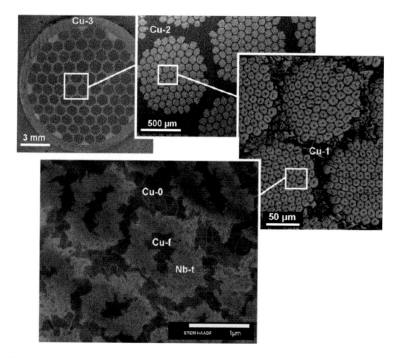

Fig. 4.15 Successive cross-sections of the multi-scale structure of the Cu/Nb/Cu nanocomposite wires. The highest magnification is a SEM image. Cu-3 is the external Cu jacket; Cu-2 and Cu-1 are Cu channels with different thickness; Cu-f is the Cu fiber inside the Nb nanotubes (Nb-t)

represents a scheme of the cross-section successive series for the nanocomposite Cu wire with Nb nanotube bundles (20.8 vol%, diameter of ~ 140 nm) alongside with their whole SEM image (adapted from [48]). The data pronouncedly demonstrate a complex, hierarchical and multi-scale character of the described Cu/Nb/Cu wire, which allows this material to be successfully used under the listed extreme actions.

4.3   Some Theoretical Approaches and Modeling

Some theoretical approaches, concerning nanotwinned structure deformation and the general pictures of SPD processes, have been mentioned before [12, 35]. An interesting tutorial review on the behavior of nanostructures under high pressures was published by San-Miguel [49]. The grain growth and GBs collective migration under the NMs plastic deformation have been analyzed by Gutkin et al. [50], and various versions of stable and non-stable boundaries migration processes are shown in Fig. 4.16 (adapted from [50]).

The GG process under an uniaxial expanding action (depending on the tension σ and orientation angles ω and Ω) can be presented by convention as a sequence of

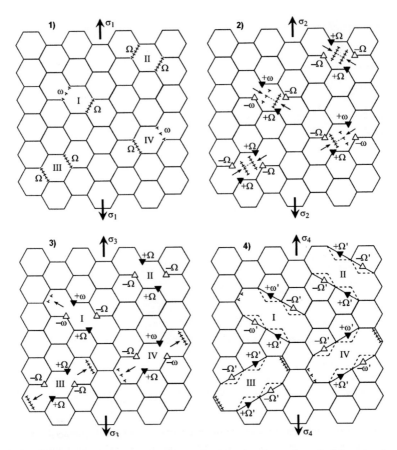

Fig. 4.16 Grain growth due to the boundary collective migration at the uniaxial tension (see the descriptions of stages 1–4 in text)

the following several stages: (1) initial state of tilt boundaries with GB angles Ω and ω at low σ_1; (2) at $\sigma_2 > \sigma_1$ the boundaries start migrating inside grains I-IV and there are appeared partial wedge-like disclinations with capacities $\pm\omega$ and $\pm\Omega$ in product double junctions; (3) at higher stress $\sigma_3 > \sigma_2$ some boundaries annihilate partly (grain I) or fully (grain II) and other pass through and stop near following boundaries (grains III and IV); 4) the boundary smoothing of effuse grains at $\sigma_4 > \sigma_3$.

The evolution of nanocracks under deformations was studied on objects with twins of both bulk and film nanostructures and described in [51]. In Fig. 4.17, various situations are shown for the nanocrack and the lamella twin ABCD (with thickness h, length d and partial dislocations at boundaries displaced at distance p from each other). The situations describe the following processes: (1) generation of a nanocrack in a grain; (2) generation of an intergranular nanocrack; (3) ABCD twin transfer into a neighbor grain.

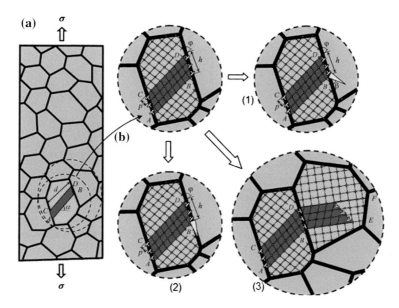

Fig. 4.17 Schematic views of the twin/nanocrack behavior in deformed NMs: **a** and **b** are *general view* and *magnified inset*, correspondingly; (*1*), (*2*) and (*3*) are different versions of behavior (see definitions in text) (adapted from [51])

The theoretical analysis applied to FCC and BCC structures has demonstrated that nanocracks in bulk NMs arise at the TBs (with width of several nm) and intensely grow with increasing the twins width h, stress σ, and angle φ. The equilibrium nanocrack length in films sufficiently decreases when the spacing between the lamella twins or their distance from the film surface is below several h values. In [52], Mg segregations were studied especially for the processes near low-angle and high-angle GBs, arising under the deformation of UFG Al–Mg alloys [52], and it was found that the boundaries segregation enrichment (earlier fixed in many experimental studies) is provoked by the stress fields of extrinsic dislocations existing near such GBs.

The results of modeling the NMs behavior under mechanical actions are presented in many publications (e.g., [53–57]) and related to various situations. For example, the modeling of fatigue crack growth resistance in NMs by MD and material mechanics methods has revealed that the crack growth processes can be retarded whenever it is possible to decrease the twins width and the distance between them to the values below 20 nm [53]. A large-scale MD simulation was carried out to evaluate the impact load action (with velocity of 600 m/s) on the nano- ($L = 10$ nm) and single-crystalline copper samples [54], for which the spall strength and the pores formation parameters were calculated depending on the nanotwins width. It was concluded that the nanotwins width reduction assist the growth of material mechanical properties only for the action on nanotwinned single-crystalline objects. The same authors, using the large-scale MD simulation, have studied the influence of

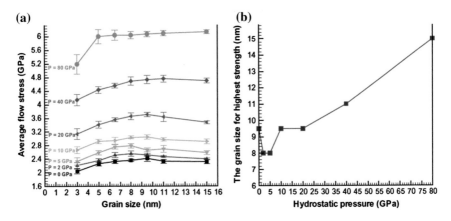

Fig. 4.18 Average flow stress of Cu as a function of the GS for different hydrostatic pressures (**a**) and the GS for highest strength versus hydrostatic pressure (**b**)

the hydrostatic pressure (up to 80 GPa) on the nanocrystal copper with a GS of 3–15 nm [55]. The results obtained are presented in Fig. 4.18, where the GS effect on the average flow stress for different hydrostatic pressures and the critical GS for the maximum strength are shown (adapted from [55]).

As it can be seen in Fig. 4.18a, the strength of Cu increases with increasing hydrostatic pressure for all investigated GS, but these dependences (despite a similar character) are placed with some shifts to each other, and therefore the hydrostatic pressure influence on the highest values is not monotonous (Fig. 4.18b). The descending and ascending character of these curves is connected with different dislocation behavior depending on both the pressure and GS as well. The growth of the intergrain boundaries thickness with the pressure rise was marked for the all range of investigated L values, and the objects with GS of 3 nm always demonstrated the tendency to amorphization under pressure of 80 GPa.

The high-temperature behavior of brittle NMs (such as diamond and boron nitride) under high pressure and plastic shear in the rotational diamond anvil cell was modeled by Levitas et al. [56, 57], where, by the finite element method, it was established that the shear deformations intensely induce phase transformations. These results manifest an important role of the deformation fields arising under the mechanical actions, and a necessity of their significance analysis for the purposes of superhard materials synthesis using shear deformations.

4.4 Examples of Applications

In addition to the above described examples (Figs. 4.14 and 4.15) of the NMs behavior and characteristics under the combined extreme conditions [46–48], it should be marked that the deformation treatment methods are widely used in many other nanotechnology fields for various purposes. In surveys [1, 58] and the publications of the 8-th International Symposium on UFG NMs (San Diego, USA, 16–20 February, 2014) [59] as well as in monograph [16], one can find the detailed consideration of material science, technology and application problems concerning many aspects of above mentioned SPD methods (see Sect. 4.1), used both for investigations practice and (less common) in commercial production as well. It is desirable to mention that the findings of Valiev and his coworkers affected the development of the SPD method: their survey [60] published in 2000 has CI over 3850.

Very interesting perspectives are opening in connection with the progress in the biomedicine implant production, particularly dental implants on the basis of Grade-4 titanium. After the SPD technology treatment, the static and dynamic strength of non-doped Ti increases by 1.5–2.5 times, allowing reducing the implant size, which leads to not only material saving, but also it is very useful for surgery operations themselves and the patient's state and health (e.g., see [61, 62]). Figure 4.19 shows a dental implant produced from nanostructured Ti (adapted from [62]).

Of other examples of using SPD methods, the improvement of the fatigue characteristics of the nickel heat-resistant HASTELLOY alloys, thanks to surface treatment by the hard-alloy balls, should be marked [63]. As shown by the SAED

Fig. 4.19 Implant from Ti (Grade-4) produced by ECAP

results, the GS at surface after treatment decreases from 30 μm down to 20 nm, the hardness rises by more than 3.5 times, and the plastically deformed zone extension is about 800 μm. After such a treatment for 30 min, the stress amplitude of the fatigue tests increases by 1.5 times (cycle number N was of 10^7). The SMGT treatment was proposed to be carried out at cryogenic temperatures [64] since such a treatment can lead to GS decreasing by 60 % as compared with the room temperature process.

For the particular problem of the production of highly wear-resistant wires on the basis of aluminum and copper, of great importance are the results concerning the HPT influence on the strength and electrical conductivity of the Al–Mg–Si [65] and Cu–Cr [66] alloys. The investigations have demonstrated that HPT at higher temperatures [65] or HPT + following annealing [66] provide the production of these alloys with the sizes of strengthening agents and particles of ~ 10 nm and ~ 200 nm, respectively. This combination also provides high values of the alloy strength, electrical conductivity and thermal stability.

Interesting examples of using SPD methods to produce Al and Mg-based superplastic metallic alloys are given in [67], where the authors using HPT method achieved the material elongation up to 400 %, which corresponds to the usual superplastic alloy characteristics. The deformation under some SPD regimes is very promising for simultaneous nanostructuring of various hybrid materials, such as metal-polymer and carbon filaments as well as spiral and helical bimetals structures [68]. In Fig. 4.20, the simplest example of a four-piece specimen is shown compressed of two Cu and two Al alloy Al6061 bits at HPT (adapted from [68]). With using the finite element simulation method, the stress fields were calculated for many spiral objects with structures of this type with a general scheme like that in Fig. 4.20.

It is interesting that these structures not only are similar to many biological objects (such as turtle shell, etc.), but also are much like they by excellent combination of some characteristics (strength, plasticity, crack propagation resistance, etc.). These layered nanocomposites are very effective in protection against bullets for body armour, micrometeorites for satellites, and high-speed particle impact for jet engine turbine blades [69]. In all these cases, there are extreme conditions for very high deformation rates at very large strains.

Fig. 4.20 The scheme of four-piece HPT combination (radius of 20 mm and thickness of 1 mm; $P = 2.5$ GPa)

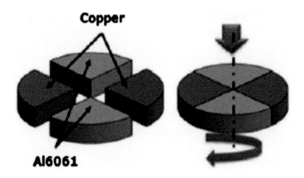

Various extreme mechanical loads are also employed in the powder technology in processes of high-energy milling, MA, different consolidation methods (pressing, rolling, hot pressing, hot extrusion, technique of high pressures and high temperatures, and plasma spark sintering), etc. Some of these methods were demonstrated above in Fig. 3.10 as related to the tube production from nanostructured ferritic steel F95. The main demand to the NMs powder consolidation implies that the maximum possible material density must be provided alongside with the minimal object porosity and the nanostructure preservation. For this, the techniques of high pressures and high temperatures as well as plasma spark sintering are the most preferential. Unfortunately, these methods are very restricted in relation to the produced objects shapes and sizes.

For the completeness of the deformation actions description, laser nanotechnology methods must be mentioned too, which include both the ablation (to produce some types of nanoparticles and films) and the creation of surfaces with improved wetting properties and gradient nanostructures. Information about the physics and technology of these processes is presented in [70–74]. It should also be noted that the significant positive effect of the gradient nanostructured surface layers on fatigue properties and strength/ductility combination was observed with respect to samples of Cu, 7075 aluminium alloy and 316/316L stainless steel processed with SMGT, warm laser shock peening, ultrasonic nanocrystal surface modification, and SRT, correspondingly [75–78]. The features formation of nanolaminated structure in nickel during SMGT were described by Lu and coworkers [79].

References

1. Estrin Y, Vinogradov A (2013) Extreme grain refinement by severe plastic deformation: a weal of challenging science. Acta Mater 61:782–817
2. Liu XC, Zhang HW, Lu K (2013) Strain-induced ultrahard and ultrastable nanolaminated structure in nickel. Science 342:337–342
3. Wang Q, Yin Y, Sun Q et al (2014) Gradient nano microstructure and its formation in pure titanium produced by surface rolling treatment. J Mater Res 29:569–577
4. Semenova IP, Salimgareeva GKh, Latysh VV et al (2009) Enhanced fatigue strength of commercially pure Ti processed by severe plastic deformation. Mater Sci Eng A 503:92–95
5. Moskalenko VA, Betekhtin VI, Kardashev BK et al (2014) Mechanical properties and structural features of nanocrystalline titanium produced cryorolling. Phys Sol State 56:1590–1596
6. Betekhtin VI, Kolobov YuR, Sklenicka V et al (2015) Defect structure effect on static and prolonged strength of submicrocrystalline Ti (BT1–0) produced by plastic deformation at screw and sizing lengthwise rolling. Techn Phys 60:66–71
7. Liu XC, Zhang HW, Lu K (2015) Formation of nanolaminated structure in an interstitial-free steel. Scr Mater 95:54–57
8. Andrievski RA, Glezer AM (2009) Strength of nanostructures. Phys-Usp 52:315–334
9. Koch CC, Ovid'ko IA, Seal S et al (2007) Structural nanocrystalline materials: fundamentals and applications. Cambridge University Press, Cambridge

10. Armstrong RW (2013) Hall-Petch analysis of dislocation pileups in thin material layers and in nanocrystals. J Mater Res 28:1793–1798
11. Lu K, Lu L, Suresh S (2009) Strengthening materials by engineering coherent internal boundaries at the nanoscale. Science 324:349–352
12. Morozov NF, Ovid'ko IA, Skiba NV (2014) Plastic flow through widening of nanoscale twins in ultra-grained metallic materials with nanotwinned structures. Rev Adv Mater Sci 37:29–36
13. Beyerlein IJ, Mayeur JR, Zheng S et al (2014) Emergence of stable interfaces under extreme plastic deformation. Proc NAS 111:4386–4390
14. Tian Y, Xu B, Yu D et al (2013) Ultrahard nanotwinned cubic boron nitride. Nature 493:385–388
15. Huang Q, Yu D, Xu B et al (2014) Nanotwinned diamond with unprecedented hardness and stability. Nature 510:250–253
16. Valiev RZ, Zhilyaev AP, Langdon TG (2014) Bulk nanostructured materials: fundamentals and applications. Wiley, Weinheim
17. Estrin Y, Vinogradov A (2010) Fatigue of light alloys with ultrafine grain structure produced by severe plastic deformation: an overview. Int J Fatig 32:898–907
18. Li RH, Zhang ZJ, Zhang P et al (2013) Improved fatigue properties of ultrafine-grained copper under cyclic torsion loading. Acta Mater 61:2857–2868
19. Dobatkin SV, Terent'ev VF, Skrotzki W et al (2012) Structure and fatigue properties of 08Kh18N10T steel after equal-channel angular pressing and heating. Russ Metall (Metally) 11:954–962
20. Zhang P, Zhang ZJ, Li LL et al (2012) Twin boundary: stronger or weaker interface to resist fatigue cracking? Scr Mater 66:854–859
21. Zhang GP, Sun KH, Zhang B et al (2008) Tensile and fatigue strength of ultrathin copper films. Mater Sci Eng A 483–484:387–390
22. Ensslen Ch, Kraft O, Mönig R et al (2014) Mechanical annealing of Cu–Si nanowires during high-cycle fatigue. MRS Commun 4:83–87
23. Dai CY, Zhang B, Xu J et al (2013) On size effects on fatigue properties of metal foils at micrometer scales. Mater Sci Eng A 575:217–222
24. Naimark OB, Plekhov OA, Betekhtin VI et al (2014) The defect accumulation kinetics and duality of Wellers' curve in gigacycle fatigue of metals. Techn Phys 59:398–401
25. Lukáš P, Kunz L, Navrátilová L et al (2011) Fatigue damage of ultrafine-grain copper in very-high cycle fatigue region. Mater Sci Eng A 528:7036–7040
26. Choi I-Ch, Yoo B-G, Kraft O et al (2014) High-cycle fatigue behavior of Zn-22 %Al alloy processed by high-pressure torsion. Mater Eng Sci A 618:37–40
27. Glezer AM, Blinova EN, Poznyakov VA et al (2003) Martensite transformation in nanoparticles and nanomaterials. J Nanopart Res 5:551–560
28. Clark SM, Prilliman SG, Erdonmez CK et al (2005) Size dependence of the pressure-induced γ to α structural phase transition in iron oxide nanocrystals. Nanotechnogy 16:2813–2818
29. Edalati K, Daio T, Arita M et al (2014) High-pressure torsion of titanium at cryogenic and room temperatures: grain size effect on allotropic phase transformation. Acta Mater 68:207–213
30. Straumal BB, Mazilkin AA, Protasova SG et al (2012) Ferromagnetism of nanostructured zinc oxide films. Phys Met Metallogr 113:1244–1256
31. Straumal BB, Dobatkin SV, Rodin AO et al (2011) Structure and properties of nanograined Fe–C alloys after severe plastic deformation. Adv Eng Mater 13:463–469
32. Straumal BB, Sauvage X, Baretzky B et al (2014) Grain boundary films in Al–Zn alloys after high pressure torsion. Scr Mater 70:59–62
33. Raabe D, Choi P-P, Li Y et al (2010) Metallic composites processed via extreme deformation: toward the limits of strength in bulk materials. MRS Bull 35:982–991
34. Sundeev RV, Glezer AM, Shalimova AV (2014) Structural and phase transitions in the amorphous and nanocrystalline $Ti_{50}Ni_{25}Cu_{25}$ alloys upon high-pressure torsion. Mater Lett 133:32–34

35. Glezer AM, Metlov LS (2010) Physics of megaplastic (severe) deformation in solids. Phys Sol State 52:1162–1169
36. Molodets AM, Golyshev AA, Shul'ga YuM (2013) Polymorphic transformations in nanostructured anatase (TiO_2) under high-pressure shock compression. Tech Phys 58: 1029–1033
37. Swamy V, Kuznetsov A, Dubrovinsky LS et al (2006) Size-dependent pressure-induced amorphization in nanoscale TiO_2. Phys Rev Lett 95:135702 (1–4)
38. Singh M, Kao M (2013) Study of nanomaterials under high pressure. Adv Nanopart 2:350–357
39. Glezer AM, Varjukhin VN, Tomchuk AA et al (2014) Basic patterns of the generation of high-angle grain boundaries and the physical/mechanical properties of Fe–Ni alloys upon severe plastic deformation. Bull Russ Acad Sci Phys 78:1022–1025
40. Han SZ, Goto M. Ahn J-H et al (2014) Grain growth in ultrafine grain sized copper during cyclic deformation. J All Comp 615:5587–5589
41. Andrievski RA (2009) Brittle nanomaterials: superhardness and superplasticity. Bull Russ Acad Sci Phys 73:1222–1226
42. Andrievski RA, Kalinnikov GV, Shtansky DV (2000) High-resolution transmission and scanning electron microscopy of nanostructured boride/nitride films. Phys Sol State 42:760–766
43. Schuster B (2011) Oxide ceramics under extreme pressure and radiation conditions. Dr. Dissertation in the Darmstadt Technical University, 2011
44. Schuster B, Fujara F, Merk B et al (2012) Response behavior of ZrO_2 under swift heavy ion irradiation with and without external pressure. Nucl Instr Meth Phys Res B 277:45–52
45. Fox-Rabinovich GS, Endrino JL, Aguirre MH et al (2012) Mechanism of adaptability for the nano-structured TiAlDrSiYN-based hard physical vapor deposition coatings under extreme frictional conditions. J Appl Phys 111:064306 (1–12)
46. Wang YX, Zhang S (2014) Toward hard yet tough ceramic coatings. Surf Coat Techn 258:1–16
47. Misra A, Thilly L (2010) Structural metals at extremes. MRS Bull 35:965–972
48. Dubois J-B, Thilly L, Renault P-O et al (2012) Cu–Nb nanocomposites wires processed by severe plastic deformation: effects of multi-scale microstructure and internal stresses on elastic-plastic properties. Adv Eng Mater 14:998–1003
49. San-Miguel A (2006) Nanomaterials under high-pressure. Chem Soc Rev 35:876–889
50. Gutkin MYu, Mikaelyan KN, Ovid'ko IA (2008) Grain growth and collective migration of grain boundaries plastic deformation of nanocrystalline materials. Phys Sol State 50:1266–1279
51. Ovid'ko IA, Sheinerman AG (2012) Nanoscale cracks at deformation twins stopped by grain boundaries in bulk and thin-film materials with nanocrystalline and ultra-grained structures. J Phys D Appl Phys 47:015307 (1–10)
52. Ovid'ko IA, Sheinerman AG, Valiev RZ (2014) Mg segregations at and near deformation-distorted grain boundaries in ultrafine-grained Al–Mg alloys. J Mater Sci 49:6682–6688
53. Chowdhury PB, Sehitoglu H, Raiteick RG et al (2013) Modeling fatigue crack growth resistance of nanocrystalline alloys. Acta Mater 61:2531–2547
54. Yuan F, Wu X (2012) Shock response of nanotwinned copper from large-scale molecular dynamic simulations. Phys Rev B 86:134108 (1–10)
55. Yuan F, Wu X (2014) Hydrostatic pressure effects on deformation mechanisms of nanocrystalline FCC metals. Comp Mater Sci 85:8–15
56. Levitas VI, Javanbakht M (2014) Phase transformations in nanograins materials under high pressure and plastic shear: nanoscale mechanisms. Nanoscale 6:162–166
57. Feng B, Levitas VI, Ma Y (2014) Strain-induced phase transformation under compression in a diamond anvil cell: simulation of a sample and gasket. J Appl Phys 115:163509 (1–14)
58. Langdon TG (2013) Twenty-five years of ultrafine-grained materials: achieving exceptional properties through grain refinement. Acta Mater 61:7035–7059

59. Mathaudhu SN, Estrin Y, Horita Z et al (2014) Preface to the special issue onultrafine-grained materials. J Mater Sci 49:6485–6486
60. Valiev RZ, Islamgaliev RK, Alexandrov IV (2000) Bulk nanostructured materials from severe plastic deformation. Progress Mater Sci 45:103–189
61. Ivanov MB, Kolobov YuR, Golosov EV et al (2011) Mechanical properties of mass-produced nanostructured titanium. Nanotechnol Russ 6:370–378
62. Mishnaevsky L Jr, Levashov E, Valiev RZ et al (2014) Nanostructured titanium-based materials for medical implants: modeling and development. Mater Sci Eng R 81:1–19
63. Dai K, Shaw L (2008) Analysis of fatigue resistance improvements via surface severe plastic deformation. Int J Fatigue 30:1398–1408
64. Darling KA, Tschopp MA, Roberts AJ et al (2013) Enhancing grain refinement in polycrystalline materials using surface mechanical attrition treatment at cryogenic temperatures. Scr Mater 69:461–464
65. Valiev RZ, Murashkin MYu, Sabirov I et al (2014) A nanostructural design to produce high-strength Al alloys with enhanced electrical conductivity. Scr Mater 76:13–16
66. Islamgaliev RK, Nesterov KM, Bourgon J et al (2014) Nanostructured Cu–Cr alloy with high strength and electrical conductivity. J Appl Phys 115:194301 (1–4)
67. Kawasaki M, Langdon TG (2014) Review: achieving superplasticity in metals processed by high-pressure torsion. J Mater Sci 49:6487–6496
68. Bouaziz O, Kim HS, Estrin Y (2013) Architecturing of metal-based composites with concurrent nanostructuring: a new paradigm of material design. Adv Eng Mater 15:336–340
69. Lee J-H, Veysset D, Singer JP et al (2012) High strain rate deformation of layered nanocomposites. Nature Comm 3:1164 (1–7)
70. De Hosson JThM, Ocelic V, De Oliveira UOB et al (2009) Fundamental and applied aspects of laser surface engineering. Int J Mater Res 100:1343–1360
71. Makarov GN (2013) Laser application in nanotechnology: nanofabrication using laser ablation and laser nanolithography. Phys-Usp 56:643–682
72. Zhang K, Dai J, Wu W et al (2015) Development of a high magnetic field assisted pulsed laser deposition system, Rev Sci Instr 86:095105
73. Khomich VYu, Shmakov VA (2015) Mechanisms and models of direct laser nanostructuring of materials. Phys-Usp 58(5). doi:10.3367/UFNr.0185.201505c.0489
74. Kolobov YR, Smolyakova MY, Kolobova AY et al (2014) Superhydrophylic textures fabricated by femtosecond laser pulses on sub-micro- and nano-crystalline titanium surfaces. Laser Phys Lett 11:125602
75. Yang L, Tao NR, Lu K et al (2013) Enhanced fatigue resistance of Cu with gradient nanograined surface layer. Scr Mater 68:801–804
76. Ye Ch, Liao Y, Suslov S et al (2014) Ultrahigh dense and gradient nano-precipitates generated by warm laser shock peening for combination of high strength and ductility. Mater Sci Eng A 609:195–203
77. Ye Ch, Talang A, Gill AS et al (2014) Gradient nanostructure and residual stresses induced by ultrasonic nanocrystal surface modification in 303 austenitic stainless steel for high strength and high ductility. Mater Sci Eng A 613:274–288
78. Huang HW, Wang ZB, Lu J et al (2015) Fatigue behaviors of AISI 316L stainless steel with gradient nanostructured layer. Acta Mater 87:150–160
79. XC Liu, Zhang HW, Lu K (2015) Formation of nano-laminated structure in nickel by means of surface mechanical grinding treatment. Acta Mater 96:24–36

Chapter 5
Nanomaterials Behavior in Corrosion Environments

Abstract The main aspects of the materials corrosion are considered taking special attention to some specific material behavior associated with the nanostructure. The NMs properties and behavior in the various corrosion media are generalized including the combined actions and high-temperature oxidation. Attention is paid to the role of size factor in reactions of nanostructures with an environment as well as to the theoretical approaches and modeling by MD methods. Some examples of the NMs exploitation in the corrosive media are given and several poorly understood phenomena are mentioned.

5.1 General Considerations

In practice, almost any use of materials relates to the interaction with some environment and it always requires accounting and prevention of corrosion. The corrosive interactions problem is of particular importance for NMs due to their non-equilibrium state. From a general view point, the presence of large number of interfaces (such as GBs and TJs), on one hand, must affect on the materials corrosion performance, because these sites (with their disordered structure) are subjected to the selective interactions with an aggressive media. On the other hand, the diffusion-controlled formation of the protective surface films (i.e., surface passivation) at these interfaces can proceed more intensively that can prevent the further spread of corrosion. General effect of some material interaction with an environment is determined by relation of these two factors and depends on reaction parameters (their kinetic rate constants), as well as on the nanostructure parameters. In this situation, the NMs can be both more or lesser corrosion-resistant in comparison with their CG counterparts, and the non-trivial situation has led to many conflicting results, related to the NMs corrosive and electrochemical behavior, that was especially noted in monograph [1] and some recent reviews (e.g., [2–7]).

© Springer International Publishing Switzerland 2016
R.A. Andrievski and A.V. Khatchoyan, *Nanomaterials in Extreme Environments*,
Springer Series in Materials Science 230, DOI 10.1007/978-3-319-25331-2_5

It seems useful to remind how the volume fractions of the GBs, TJs, and total interface area depend on the crystallite sizes. These dependences shown in Fig. 5.1 (adapted from [8]) have been derived in a frame of some model, where the grains were approximated by tetrahedral dodecahedrons divided by the intergrain boundaries of 1 nm thick.

The curves in Fig. 5.1 show that the contribution of interfaces becomes substantial when the GS is below 60–70 nm, but the TJs fraction is prevailing for the GS less than 3 nm. The total fraction of interfaces achieves 50 % for GS about 6 nm.

For example, consider the opposite influence of the GS on the oxidation resistance and compare the behavior of Fe–Cr alloy and nickel under oxidation. The curves presented in Fig. 5.2 illustrate that the transition to nanostructure is

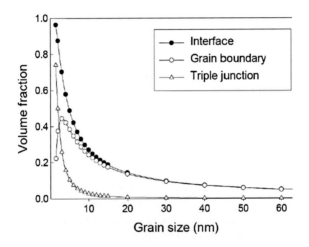

Fig. 5.1 Volume fractions of total interfaces, GBs, and TJs as a function of GS

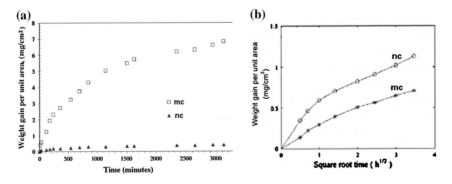

Fig. 5.2 Oxidation kinetics of nanocrystalline (*nc*) and microcrystalline (*mc*) specimens of Fe–10 wt% Cr (*T* = 400 °C) (**a**) and Ni (*T* = 700 °C) (**b**)

accompanied with the oxidation rate decrease for alloy, but the opposite tendency has been revealed for pure nickel (adapted from [9, 10]).

The emphasis of the studies [9, 10] was on the sample attestation and preparation by sintering of the Fe–Cr nanopowders and electrodeposition of Ni. The GS in the studied samples were of 52 ± 4 nm (nanocrystals) and about 1.5 μm (microcrystal) for Fe–Cr alloy, as well as of ~ 28 nm (nanocrystals) and about 3.4 μm (micro-crystals) for Ni. A comparison with data of Fig. 5.1 is demonstrating that the GS of the samples used in [9, 10] were quite representative ones. The secondary ion spectroscopy measurements have shown that the chromium and oxygen content in the surface layers of the oxidized nanostructured Fe–Cr alloys is higher than that in the microcrystalline objects. This fact confirmed a conclusion that, in these systems, a passivating oxide film formation (preventing the nanocrystalline samples corro-sion) proceeds more intensively as exemplified by the kinetic curves of Fig. 5.2a. The analogous estimations of a prevailing total diffusion flux (including the Cr partial coefficients of volume/boundary diffusion) were presented in review [11]. It is important that the lowest values of oxidation rates for the Fe–Cr nanoalloys are retained in the samples with a bimodal structure (nc + mc = 50/50) usually char-acterized by the high strength and plasticity [12]. For the nanocrystalline nickel samples, it was demonstrated that the accelerated NiO film formation is also explainable with Ni boundary diffusion, but this film is not passivating one.

The above given examples readily illustrate an important role of the GS factor on the NMs oxidation process, but for general consideration, it is necessary to keep in mind a very complicated character of the corrosion effects in a whole, because they include, for example, the pitting (i.e., formations of the pointed or localized cor-rosion sites), corrosion stress cracking, electrochemical interactions, the evolutional changes of the passivating films, galvanic pairs formation, and many other complex processes. The bibliography of studies in this field is very extensive, and therefore, considering the NMs behavior under extremes, only the most reliable and certain experimental results, obtained for the well attested samples, will be referred to. The more detailed information can be found in reviews [2–7].

In liquid media, the corrosion is studied mainly by the electrochemical methods using so- called anode and cathode characteristics curves of the $E = f\,(\lg i_a)$ type, where E is a potential and i_a is a velocity of anodic reaction, expressed by units of current density. From these characteristics, the main corrosion process parameters (such as the corrosion and passivation potentials, corresponding currents, etc.) can be determined and analyzed. The linear part of so- called the Tafel's curves is always used for the corrosion process rates determination and has a simple form:

$$E = a + b \lg i_a, \tag{5.1}$$

where a and b are some constants.

5.2 Main Experimental Results

5.2.1 Metals and Alloys

5.2.1.1 Nickel, Cobalt and Cupper

Some episodic studies of the NMs corrosion properties and special characteristics began more than 30 years ago. The first systematic study has been devoted to the electrochemical behavior of the nanocrystalline nickel samples of very high purity (99.99 %) in 2N H_2SO_4 at 293 K [13], but it is of interest that the results were published only in 1991 year. The study itself has been devoted to a comprehensive investigation of the electrodeposited nickel tapes having the high physical and mechanical characteristics [8]. In Fig. 5.3, the anodic polarization curves for nano/CG samples are presented and some passivation characteristics derived from these curves are given in Table 5.1 (adapted from [13]).

The data demonstrate that the transition to nanostructure is accompanied by an increase the corrosion and passivation potentials (E_{corr} and E_p) characteristic values. The rising of two other important parameters (namely, the critical passivation current i_a and passive state current i_p densities) shows that the electrochemical

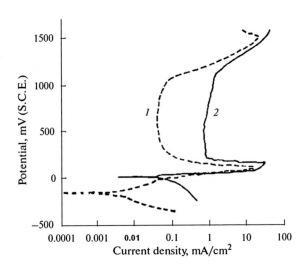

Fig. 5.3 Potentiodynamic anodic polarization curves for Ni in poly- and nanocrystalline state (*1* GS of 100 μm; *2* GS of 32 nm) in 2N H_2SO_4 at 293 K. A saturated calomel electrode was used as a reference electrode

Table 5.1 Passivation characteristics of nickel specimens with different GS (*L*) according to the data of polarization measurements in 2N H_2SO_4

L (nm)	Corrosion potential E_{corr} (mV)	Passivation potential E_p (mV)	Peak active current density i_a ($\mu A/cm^2$)	Minimum passive current density i_p ($\mu A/cm^2$)
10^6	−170.0	+113	13,370	32.7
32	+7	+158	26,120	580

resistance of nanocrystalline nickel in 2N H_2SO_4 is lower in comparison with the CG nickel as similar to its oxidation resistance (see Fig. 5.2b). The SEM studies have shown that the GBs and TJs are markedly dissolved under such conditions but the corrosion localizations were higher for the CG samples. It was found also that nanostructurization of the nickel samples is catalyzing the hydrogen impurities release as well as retarding the passivation kinetics and decreasing a stability of the formed passivating films. The range of the studied GS was later widened [14] to the smaller values (L = 28, 22, 10 and 8 nm) and the results obtained have largely confirmed the main results of [13], namely, a growth of E_p and i_p values with the GS decreasing. It was marked in review [8] that the commercial fabrication of the studied nanocrystalline nickel bands (in spite of some decrease of their corrosive resistance) was organized in Canada and they had a widespread use for coating and cladding for the various items in power-engineering and other technical fields.

The comparison of the corrosion effects in the nano/CG cobalt samples ($L \sim 12$ nm and $L \sim 8$ μm, respectively) in 0.25M Na_2SO_4 solution (pH of 6.5) has shown that the GS decrease is only slightly influencing on the material corrosion resistance [15]. However, for the copper electrochemical corrosion in 0.1M $HNO_{3,}$ a transition to the nanostructure (namely from GS of ~ 300 μm to L about 20 nm) is accompanied by more active dissolution without any passivation, that can be explained by a presence of numerous GBs, TJs, microvoids, and weak links at certain interfaces between grains [16]. In addition, for the general analyze of the GS influence on the material corrosion resistance, the considerations must be given to many other factors, such as the residual stresses, the impurity localization (for example, sulfur) on the grain boundaries, the surface roughness and texture, the pores presence, the passivating layers stability, etc. These numerous and various material features often lead to the non-monotonic characteristics of the corrosion resistance under a grain refinement or to a reduction of the real size effects [3, 7, 17].

In connection with a diversity of these influencing factors, it is of interest to consider the results of the Cu foils behavior (with both the nanotwinned and microcrystalline structure) in 3.5 % NaCl solution (pH $\sim$ 8.0), presented in Table 5.2 (adapted from [18]).

Table 5.2 Summary of corrosion test of nanotwinned and microcrystalline Cu foils with common thickness of ~ 25 μm

Type	Grain size (μm)	% grains with with na- notwins	Area fraction of grain {111} orientation	Polarization resistance (kΩ cm^2)	Corrosion cur-rent density (μA cm^{-2})	Passivation current density (mA cm^{-2})
I	0.52	90	0.88	28.6 ± 7.1	0.34 ± 0.09	1.15 ± 0.20
II	0.41	60	0.33	10.2 ± 4.4	1.08 ± 0.39	1.14 ± 0.23
III	0.34	5	0.08	10.4 ± 0.5	1.26 ± 0.06	1.08 ± 0.27
IV	8.6	0	0.02	9.2 ± 2.1	1.92 ± 0.42	1.27 ± 0.18

The data readily manifest that namely the films with high concentration of the nanotwinned grains and their area fraction with the {111}-texture demonstrated the highest corrosion resistance though the passivation current densities values were the same for all objects. The investigations of the protective passivating layers content (including both the electrochemical corrosion studies and immersion experiments) detected a presence of the Cu_2O columnar (bar) crystals with the {111}-texture, but a minimal pitting was observed in the type I foils. It is interesting that the corrosion characteristics of III and IV type foils (in spite of the GS difference about one order of magnitude) are very similar, i.e., in the experiments [18] the GS factor was not an important one.

5.2.1.2 Iron and Steels

These objects are well studied (e.g., papers [9, 12, 19–23], reviews [1–7, 11] and references therein), and therefore we will give attention to the last results in brief. A wide investigation of electrochemical properties of the α-Fe + Fe_3C nanocrystalline composites in some sulfuric and hydrochloric acid media was carried out by authors [19] for samples prepared by mechanosynthesis from the carbonyl Fe and carbon mixtures followed by a dynamic pressing. In the Table 5.3 (adapted from [19]), the detailed data are presented concerning the characterization and properties of numerous composites including the compact samples of pure Fe and steel with carbon content of 1.3 mass%.

The analysis of the obtained polarization curves and values of E_{corr} and i_a has shown that for the $Fe_{95}C_5$ and Fe–1.3 % C samples the transition to nanocrystalline state is markedly influencing the active dissolution process (that can be obviously explained by the GBs number growth), but from other side, annealing of the $Fe_{95}C_5$

Table 5.3 Characterization and electrochemical properties of (α-Fe + Fe_3C) composites in 0.05M H_2SO_4 + 0.45M Na2SO4

Composite	Phase content (wt%)		Grain size (nm)		Parameters of polarization curves			
	α-Fe	Fe_3C	α-Fe	Fe_3C	E_{corr} (mV)	i_c(−400 mV) (mA/cm^2)	i_a (−200 mV) (mA/cm^2)	i_p (1250 mV) (mA/cm^2)
Pure Fe	100	–	$300 \cdot 10^4$	–	−325	0.6	35.20	0.05
Fe–1.3 %C	~91	~9	$2 \cdot 10^4$	n/d	−305	1.6	45.2	0.1
$Fe_{95}C_5$	91	9	38	48	−385	1.1	65.4	1.4
Annealing at 800 °C	91	9	n/d	n/d	−305	5.7	48.1	1.8
$Fe_{90}C_{10}$	84	16	40	49	−385	1.8	99.7	4.5
$Fe_{85}C_{15}$	45	55	49	47	−325	7.8	26.2	1.0
$Fe_{75}C_{25}$	8	92	42	29	−195	26.2	~0	3.9

E_{corr} is a corrosion potential; i_c, i_a and i_p are cathodic, anodic and passivation current densities at different potentials, respectively

samples, leading to the GS growth, is leveling the difference with the Fe–1.3 % C steel regarding the values of E_{corr} and i_a. The microscopic studies have shown also an important role of a forming cementite morphology, because the protective role of the γ-FeOOH and Fe_2O_3 passivating films is decreasing due to carbon accumulation at the interfaces resulting from the cementite decomposition. The authors [18] concluded that namely the cementite (Fe_3C) content makes the greatest impact on the electrochemical behavior of the studied composites and this influence exhibits itself not only in the anode and cathode current values, but also in the growth of the pitting resistance and a catalytic activity in the hydrogen release reaction.

In some works, a positive effect of nanocrystallinity on corrosion resistance of the pure iron and its chromium alloys has been demonstrated. For example, in [22] there has been shown, that the electrochemical resistance of the nanocrystal Fe samples ($L = 50$–89 nm) in the (0.1–0.4)M HCl solutions is increasing as compared with the CG counterparts ($L \sim 50$ μm). The similar results (with the GS decrease from 750 to 32 nm) were obtained for the Fe electrochemical corrosion in near-neutral aqueous solution containing corrosion inhibitor like sodium benzoate [20]. A transition to the nanostructures for the Fe–(10, 20)% Cr alloys was accompanied with more active passivation and formation of a protective Cr_2O_3 film, leading to improvement of the electrochemical corrosion resistance in 0.5M H_2SO_4 [23], that qualitatively corresponds to the above given results (Fig. 5.2a) for these alloys oxidation [9].

5.2.1.3 Titanium and Zirconium

In a CG state, these two metals under aggressive environment are coated by some oxide protective films impeding any corrosion development. Therefore, these metals transition to a nanocrystalline state is accompanied with various phenomena, connected with passivation processes, which were a subject of many investigations (e.g., [24–31]). The comparison of the electrochemical behavior for both nanostructured and CG samples demonstrates an important role of the Ti texture (arising at different modes of ECAP and HPT actions) and a non-monotonic influence of the GS. Such non-monotonic kinetic changes for the corrosion resistance of microscopic samples were observed also for Ti and Ti–V–Al alloys, as well as corrosion resistance rise for alloys after ECAP [26, 27]. The detailed studies of the electrochemical properties and corrosion resistance of the CG ($L \sim 15$ μm) and submicrocrystalline ($L \sim 0.15$ and 0.46 μm) samples in the (1–5)M H_2SO_4 have demonstrated some important differences in the anode and cathode process rates and other parameters [28]. In Fig. 5.4, the kinetic curves of the mass change per unite surface area are presented for different titanium samples in 5M H_2SO_4 at 23 °C (adapted from [28]).

As it can be seen from the Fig. 5.4, an induction period is a characteristic for the microcrystalline samples only, but in one of them (the sample with smallest GS, curve 3) the corrosion instead of decay becomes very intense and dominates in the

Fig. 5.4 Time dependence of the relative change in weight upon corrosion in 5M H_2SO_4 of Ti specimens with different GS: (*1*) $L = 15$ μm; (*2*) $L = 0.46$ μm; (*3*) $L = 0.15$ μm

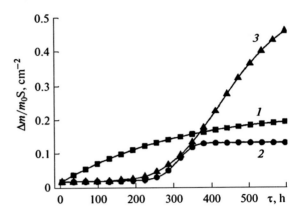

Table 5.4 Corrosion current density (i_{corr}) and corrosion rate (on the weight loss results after 720 h duration) of different $Ti_{60}Ni_{40}$ specimens in 1M NaCl

Specimen	Grain size (nm)	i_{corr} (mA/cm^2)	Corrosion rate (mm/year)
Nanocrystalline multi-phase (Ti_2Ni + Ti + TiNi)	35 ± 5	$1.5 \cdot 10^{-3}$	0.066
Amorphous ($Ti_{60}Ni_{40}$)	–	$8.3 \cdot 10^{-4}$	0.029
Nanocrystalline (Ti_2Ni)	10 ± 5	$4.0 \cdot 10^{-4}$	0.017

GS sample. The mechanism of this phenomenon is interesting and needs new research efforts.

The thermographic studies of oxidation these objects have shown that in the polythermal mode up to $T \sim 1000\ °C$, the weight changes are practically the same for all samples, and only at higher temperatures the microcrystalline samples are oxidizing more intensively [29]. It was found also that the microstructure, formed in Ti under deformation effects, promotes more active oxygen dissolution and stabilization of α-phase by shifting the α → β polymorphic transition point to the high temperature range.

The electrochemical behavior and immersion corrosion of the nanocrystalline and amorphous intermetallic $Ti_{60}Ni_{40}$ samples in 1M NaCl aqueous medium ($T = 20\ °C$) were studied and compared in [30]. The data are presented in Table 5.4 (adapted from [30]).

The results show that the substantially more positive results in the both corrosion tests were obtained for the single phase nanocrystalline samples. The XPS studies have demonstrated that only in this case the TiO_2 film is formed over the sample surface (in other samples their surfaces contained the TiO and Ti_2O_3 oxide admixtures as well as metallic titanium).

In [31], the ECAP (4 passes at $T = 460\ °C$) influence was detected for one of the most important material characteristics (namely, stress corrosion cracking) by the example of Zr–2.5 wt% Nb alloy widely used in the nuclear technique. As the

Table 5.5 Structure parameters and the strength/elongation change of Zr–2.5 wt% Nb alloy after the stress corrosion cracking tests

Specimen treatment	Structure parameters: L is a GS, N_{pitt} is a number of pittings, and d_{pitt} is a pitting diameter			Temporal fracture resistance (σ_t) and elongation (δ) before (numerator) and after (denominator) tests			
	L (nm)	N_{pitt}	d_{pitt} (nm)	σ_t (MPa)	$\Delta\sigma_t/\sigma_t$ (%)	δ (%)	$\Delta\delta/\delta$ (%)
ECAP	200–700	115 ± 12	5 ± 2	700/665	5	10/9	10
Annealing (1)	400–1100	60 ± 6	16 ± 2	680/545	20	15/9	40
Annealing (2)	500–1500	9 ± 5	14 ± 2	620/510	18	25/17	32

corrosive medium, 1 % iodine solution in methanol was used, the stress level corresponds to 0.8 of the yield strength limit, the temperature equals to 20 ± 2 °C and the test duration was 50 h. For comparison, the samples annealed by two modes were used: (1) 530 °C (1 h) and (2) 530 °C (1 h) + 560 °C (3 h). The sample characterization and test results are presented in Table 5.5 (adapted from [31]).

From these data, we notice that the losses in mechanical properties are significantly lower for the samples treated by ECAP. However, the results are not solid, because of numerous nanosized pitting defects were detected in these samples, which under durable tests can be converted into the microscopic cracks or fracture sites, that maybe will cardinally change the whole situation of the described mechanical properties degradation after the samples testing.

5.2.1.4 Aluminum, Magnesium and Zink

It is interesting to note that earlier the specialists (for example, authors of review [2]), when considered some inconsistency of the results concerning the GS influence on a material corrosion behavior (in Al, Mg and their alloys), paid their main attention to the systems with micrometric GS, but lately the interest has been clearly shifted to the so-called ultrafine-grained (UFG) materials. Nevertheless, the nanostructurization of low-melting metals by SPD methods as before remains connected with a complex and hard problem, because of the intensive dynamic recrystallization. The electrochemical tests of Al–4 wt% Mg–0.8 % Si alloy samples in sea water have discovered that the GS refinement from of 7 μm to about 0.4 μm leads to the polarization resistance growth about 1.5 times [32], but at the same time, the stress corrosion cracking turned to be substantially below for the CG samples. After 4 passes of ECAP, the GS of AA2024-T351(Al–4.15 % Cu–1.14 % Mg) alloy decreases approximately to 200 μm and it manifests itself during transition from an intercrystalline corrosion mode to a pitting one under tests in 0.5 M NaCl solution [33]. In samples of the magnesium AZ31 alloy (Mg–3.62 % Al–1.36 % Zn), the GS achieved about 1 μm after 8 passes of ECAP. The ECAP treated samples were characterized by a greater uniformity and homogeneity, and their corrosion resistance in 0.1 M NaCl solution (exposure time is of 168 h) was higher as compared with the initial CG samples [34].

The increase of the electrochemical corrosion resistance with the GS diminishing in the range from of 5–20 μm to about 40–60 nm was detected also for the electrodeposited micro- and nanostructured zinc coatings [35, 36].

5.2.2   High-Melting Point Compounds

Conventionally, a HMPC group includes carbides, borides, nitrides, oxides and other substances with the melting point temperature T_M above 2000 °C. The thermal, radiation and deformation stability of such nanocrystalline objects was considered in early Chaps. 2–4. Below, we analyze a behavior of such materials in some liquid media and under a high-temperature oxidation.

5.2.2.1   Behavior in Liquid Media

In general, the problem of HMPC corrosion (including the high-temperature oxidation processes) was studied by Lavrenko and his coworkers (e.g., [37–41]). As an example, in Fig. 5.5, the anode polarization curves are presented for the amorphous and amorphous-nanocrystalline TiB_2 films as well as CG hot-pressed samples tested in sea water at room temperature.

It was supposed [40] that the TiB_2 anode oxidation can be described by the following system of reactions:

$$TiB_2 + 7H_2O \rightarrow TiO^{2+} + 2BO_3^{3-} + 14H^+ + 10e, \qquad (5.2)$$

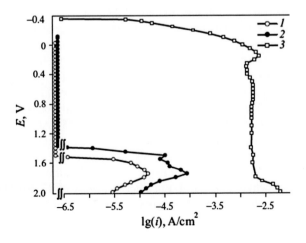

Fig. 5.5 Anodic polarization curves of TiB_2 amorphous (*1*), amorphous-nanocrystalline (*2*) and CG (*3*) specimens (adapted from [40])

$$TiB_2 + 6H_2O \rightarrow Ti^{3+} + 2BO_3^{3-} + 12H^+ + 9e, \qquad (5.3)$$

$$TiB_2 + 8H_2O \rightarrow TiO_2 + 2BO_3^{3-} + 16H^+ + 10e, \qquad (5.4)$$

where the first reaction (5.2) corresponds to the more sloping parts of the polarization curves (1 and 2) and reflects a partial electrochemical dissolution of titanium diboride. The steeper part of these curves corresponds to the trivalent titanium ions transitions into solution by the reaction (5.3) and the final passivation process is connected with the TiO_2 (rutile modification) formation by the reaction (5.4). The results of analysis for the properties (before and after the electrochemical testing) allow to give several common conclusions:

1. Corrosion resistance of the TiB_2 CG samples is sufficiently low as compared with their film counterparts.
2. Corrosion resistance of amorphous films is one order of magnitude higher than that for the amorphous-crystalline films containing various boride phases (such as TiB_2, TiB and other compounds) with GS in the 15–90 nm range.
3. In the studied film thickness range (70–250 nm), the corrosion resistance and passivation potential of the amorphous films increase with their thickness.
4. The films oxidation proceeds by the pitting corrosion mechanisms and the TiO_2 formation is fixed inside the pittings. Such pitting-type character of corrosion for the amorphous-crystalline TiB_2 films and their high corrosion resistance were described also in [42].

The similar and analogous results were obtained also in studies of the TiN films electrochemical corrosion [41], and an essential growth of the corrosion resistance was fixed in many other investigations of the nitride films behavior in liquid media of sea water type. Some recent results of these studies are presented in the following studies, devoted to the titanium nitride and alloys in its base [43–46], as well as to the zirconium, hafnium, vanadium, and chromium nitrides [47–50]. The data of Table 5.6 illustrates the various characteristics of the TiN films being deposited at different temperatures on stainless 304 steel surface by a magnetron sputtering method. These results pronouncedly demonstrate that the film thickness decrease leads to substantial growth of their corrosion resistance in 0.5M H_2SO_4 solution, and this effect can be connected with a so- called fine-grained morphology under

Table 5.6 Effect of different substrate temperature (T_S) on the TiN films thickness (δ), surface roughness (*RMS* is root-mean-square value) and corrosion characteristics (E_{corr}, i_{corr} and corrosion rate) (adapted from [44])

T_S (°C)	δ (nm)	RMS (nm)	E_{corr} (V)	i_{corr} ($\mu A/cm^2$)	Corrosion rate (mm/year)
25	205	3.95	0.433	9.1	0.0947
100	181	1.84	0.489	7.48	0.0775
200	152	1.71	0.469	3.73	0.0390
300	137	1.62	0.523	1.78	0.0181

the deposition temperature rise. It is characteristic that the corrosion resistance growth is accompanied with some increase of the materials wear resistance (e.g., [45–47, 49]).

5.2.2.2 High-Temperature Oxidation

Some results, concerning the oxidation of the HMPC films, were presented above as applied to the cutting materials in Chaps. 2 and 4 (see Figs. 2.7–2.9 and 4.14). An interest to these investigations is now rising in connection with the special materials for hypersonic aviation (the fairings, wing leading sharp edges, propulsion system components, etc.). The oxidation, degradation and corrosion behavior of many composite materials-based ZrB_2, HfB_2, SiC, $MoSi_2$, and other compounds are now investigated at the very wide temperature range (e.g., [51–53]). These materials are now called as the ultra-high temperature ceramics. It must be noted that a role of size effects (both at the micro- and nano levels) in bulk samples of these materials practically remains unexplored, and therefore below, we give some examples of the nanoparticles and films behavior at oxidation.

In Fig. 5.6 (adapted from [54]), three different exothermic peaks are very pronounced for the TiC micro- and nanoparticles (in details, they are described below in Table 5.7).

The data show that the particles of similar stoichiometry have only few differences in chemical composition and their particle sizes (evaluated by BET methods) are differ in 800 times. Nevertheless, the phase formation during oxidation (fixed by

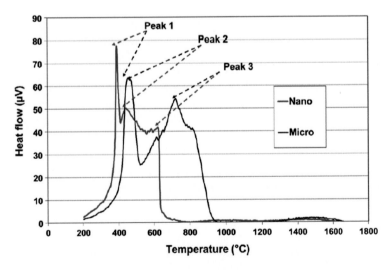

Fig. 5.6 The evolution of heat flow as a function of DTA temperature for TiC nano- and microparticles (see Table 5.7)

Table 5.7 Characterization of TiC particles and temperature range of phase occurrence during oxidation (adapted from [54])

Particle type	Formula	Diameter (nm)	Temperature range (°C) of phase occurrence during oxidation				
			TiC	TiC_xO_{1-x}	Ti_3O_5	TiO_2 (anatase)	TiO_2 (rutile)
Nano	$TiC_{0.9}O_{0.23}N_{0.10}$	53	<300	<390	330; 620	620; 800	>800
Micro	$TiC_{1.0}O_{0.04}N_{0.01}$	$4 \cdot 10^3$	<300	<460	400; 720	720; 900	>900

XRD) was a similar type with characteristic shift to the lower temperatures just for nanoparticles.

The study of the WC micro- and nanoparticles oxidation in a wide range of sizes (from about 20 to of 2000 nm) and temperature interval (323–1173 K) has shown that the WC $\rightarrow$ W_2O_3 transition (without WO_2 formation) does not depend on the particles dispersity character [55]. The oxidation rate was proportional to $1/<d>$, where $<d>$ is the average particle size, and the oxidation process activation energy was falling from value of $\sim$ 120 kJ/mol to about 90 kJ/mol with the particle size diminishing from about 2000 nm to approximately 20 nm.

The diboride oxidation is describing by several reactions from which, especially for TiB_2, we can single out the following ones:

$$3TiB_2 + 7O_2 \rightarrow TiO_2 + 2B_2O_3 + 2TiBO_3, \tag{5.5}$$

$$4TiB_2 + 9O_2 \rightarrow 4TiBO_3 + 2B_2O_3, \tag{5.6}$$

$$4TiBO_3 + O_2 \rightarrow 4TiO_2 + 2B_2O_3, \tag{5.7}$$

$$2TiB_2 + 5O_2 \rightarrow 2TiO_2 + 2B_2O_3. \tag{5.8}$$

The boron oxide and titanium borate formation usually is fixed by the XRD, DTA/TGA methods, and so on. The characteristic feature of the (5.5–5.8) reactions is the evaporation of the low-melting boron oxide, imposing a specific influence on the TiO_2 passivating films formation. The admixture role in a shift of the TiO_2 film high-temperature formation was marked also in [37] for the different in composition TiB_2 powders. The powders and compacts with a higher content of metallic admixtures (such as Fe, Ni, Mo and others) turned to be more stable to oxidation in the 700–1200 °C range and, for them, the high-temperature peak at the DTA/TGA thermograms was fixed at about 80 °C higher in comparison with their more pure counterparts. This fact can be naturally connected with the interface enrichment by the metallic admixtures followed by formation of a diffusion barrier inhibiting further oxidation process. The TEM observations in situ of the ZrB_2 nanoparticles (with size of $\sim$ 35 nm) oxidation behavior at temperature of 1500 °C have revealed the reaction of (5.8)-type with formation of a ZrO_2 tetragonal modification and evaporation of B_2O_3 [56].

Fig. 5.7 DTA (*1*) and TGA
(*2*) curves of the oxidation of
films I (**a**), II (**b**) and III (**c**)

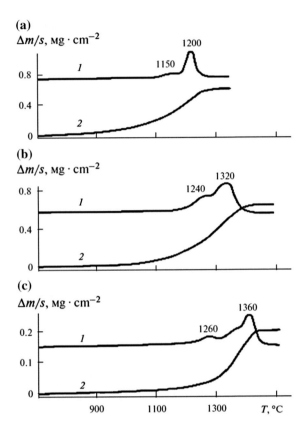

Let us consider some results of the film oxidation studies supplementary to
above presented data of Chaps. 2 and 4. In Fig. 5.7, the DTA/TGA curves are
presented for oxidation of the multicomponent films, obtained by a magnetron
sputtering using the targets with different composition (in wt%): AlN + 50TiN
(film **I**), AlN + 50TiB$_2$ (film **II**) and AlN + 10TiB$_2$ + 20SiC (film **III**) (adapted
from [38]).

These data show that, under the study heating conditions (15 °C/min), a marked
oxidation process for the films **I**, **II** and **III** is fixed at the temperature
of ∼750, ∼950 and ∼1020 °C, respectively. It is obvious also that film **III**
demonstrates the highest oxidation resistance. These data comparison with the
results for CG samples, being oxidized under the same conditions, shows that the
values of specific mass gain under oxidation for the film samples are lower in 4–5
times.

Below, in general form, is written down the main reactions proceeding under the **I–III** type films oxidation:

film **I**

$$2(Ti_{1-y}Al_y)N_x + (2 - y/2)O_2 \rightarrow 2(1 - y/2)TiO_2 + yAl_2O_3 + xN_2, \qquad (5.9)$$

$$TiO_2 + Al_2O_3 \rightarrow Al_2TiO_5; \qquad (5.10)$$

film **II**

$$2TiB_2 + 5O_2 \rightarrow 2TiO_2 + 2B_2O_3, \qquad (5.11)$$

$$4AlN + 3O_2 \rightarrow 2Al_2O_3 + 2N_2, \qquad (5.12)$$

$$2Al_2O_3 + B_2O_3 \rightarrow Al_4B_2O_9, \qquad (5.13)$$

$$9Al_2O_3 + 2B_2O_3 \rightarrow Al_{18}B_4O_{33} \qquad (5.14)$$

and reaction (5.10);
   film **III**

$$SiC + 2O_2 \rightarrow SiO_2 + CO_2, \qquad (5.15)$$

$$3Al_2O_3 + 2SiO_2 \rightarrow 3Al_2O_32SiO_2 \qquad (5.16)$$

and reactions (5.10–5.14).

In these reaction sets, no account has been taken to a possible admixture influence, such as iron (always connected with the powder samples grinding under target production), the non-stoichiometric phase formation, nitrogen reactions, etc. From the general considerations, it is obvious that the studied process of high-temperature films interaction with air is a multi-stage one. An oxidation initial stage practically for all films begins with a formation of the rutile and boron oxide layers (for films **II** and **III**). With allowance made for the known data, regarding to the oxidation processes in the CG objects with the same composition, the DTA curve peak at 1200 °C for film **I** can be associated with the reaction (5.10), but the peaks at 1240 and 1320 °C (film **II**) most probably correspond to the (5.13, 5.14 and 5.10) reactions. The peaks at 1260 and 1390 °C (for film **III**) relate to the reactions (5.15 and 5.12), connected with the β-crystobalite and α-alumina formation, while the peak at 1390 °C corresponds to the reactions (5.10 and 5.16) with formation of aluminum titanate and mullite (the last two reactions most likely define a protective layer composition for the film **III**). The material phase compositions obtained in the reactions (5.13) and (5.14), however, are partly verified only by the XRD data, maybe because the films are very thin and the total phase amount remains low for correct registration.

The results of microspectral X-ray analysis, obtained by scanning the
0.2 mm-length film surface sections, are confirmed in the films **I–III** after oxidation
a presence of some phase near by content to aluminum titanate. An EDA detected
also confirmed that the surface layers are enriched by aluminum and depleted by
titanium. The fracture surface analysis of the films has shown a formation of some
nanocolumnars with the characteristic sizes of ∼100 nm. Moreover, this phenomena was observed only in the films **III** whereas such nanocolumnars were
absent in films **I** and **II**, and for them, the structure inclusion dimensions after
oxidation were in the range 300–600 nm.

Hence, the high scaling resistance, detected for the film **III,** can be explained by
enrichment of the surface with aluminum and a stable protective layer formation. In
this case, the nanocolumnar formation is assisting to nanocrystalline structure
conservation and more intensive diffusion mobility. It maybe supposed that namely
nanostructures, as in given above results (Fig. 5.2a) [9], are ensuring a more
intensive formation of the stable protective layer.

The oxidation processes and corrosion resistance comparison for the simple and
complex HMPC films was carried out and analyzed in [57]. Figure 5.8 shows some
advantages of the amorphous films over the crystalline ones.

Though the Fig. 5.8 results relate only to the short-time experiment durations,
but in general, they seem to be correct. It is interesting that the Si–B–C–N based
films on the sapphire and silicon carbide substrates remained stable up to temperatures of 1700 °C.

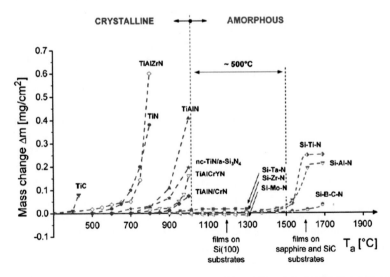

Fig. 5.8 Oxidation resistance of crystalline and amorphous films-based HMPC (adapted from
[57])

5.3   Some Theoretical Approaches and Modeling

The results of many cited studies (e.g., [9, 12, 18, 23, 27, 39–41, 45–47, 57]) are clearly demonstrating that various studied NMs really have the high corrosion and oxidation resistance characteristics. However, these results do not give a comprehensive idea of the GBs, TJs, grain inner areas or other peculiarities role in the investigated systems, and while these studies remain predominantly at the empirical level. The theoretical investigations are very few.

Some results, obtained for nanocrystalline zirconium oxidation kinetics in the 200–500 °C range and calculated with taking into account the electronic mean free pass, are presented partly in Fig. 5.9 [58]. They show a data good agreement for the CG samples and a general decrease of the zirconium oxidation rate with the decrease of GS.

A relationship between the GS and corrosion rate of metals, similar to the known formulae of Hall-Petch (see early (4.4)), was proposed in [59] and has a form:

$$i_{\text{corr}} = (A) + (B)L^{-1/2}, \tag{5.17}$$

where i_{corr} is the corrosion current density (the value proportional to corrosion rate), A is the constant which is likely proportional of the environment and B represents a material constant according composition and impurity level. This relationship is based on the experimental data extension and generalization for the CG Mg and Al alloys, but its physical foundations remain still elusive.

The approximations in the DFT method frame have shown that the TiB_2 nanocrystals are stable with respect to water molecules and thus can be reckoned as resistant to attack by the various biological media [60]. The examples of the DFT usage for to describe the passivity films breakdown are presented in review [5] as an example for a Cl interaction with hydroxylated NiO [61]. The theoretical consideration and comparison of the one- and multi-layer (as well as gradient ones)

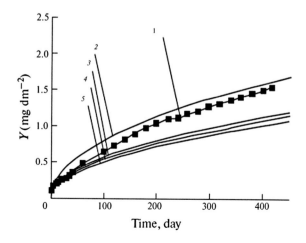

Fig. 5.9 Oxidation kinetics ($T = 400$ °C) of Zr with different GS: (1) experimental data for CG specimens; (2) estimated data for those specimens; (3, 4 and 5) estimated values for GS of 29, 25 and 20 nm, respectively (adapted from [58])

coatings are presented in [62], where it was shown a perceptivity of compositionally graded coating structures for the operational service life growth. Other theoretical approaches of the corrosion studies towards multy-scale modeling of localized corrosion are presented in review [63], but such investigations are only starting and still practically have not connection with the NMs itself (as a rule, in such studies, the size effects are not taken in mind).

5.4   Examples of Applications

There are many data in literature concerning the nanostructured coating perspectives for the corrosion attack reduction in different systems. As an example, in Fig. 5.10, the kinetic curves are shown for the mass changes of the ASTM 1020 steel samples (in initial state and with various coatings) in a sulphate media at high temperatures [64]. The coatings-based Ni–Cr–C alloy were formed by the high-velocity air-fuel deposition method, using the standard CG powders and powders, being milled at low temperatures (in the latter case the GS was about 30–50 nm). The Fig. 5.10 data are clearly demonstrate that these coatings are considerably decreasing the corrosion actions at the sacrifice of a dense Cr_2O_3 film formation, being fixed by the XRD, SEM, and TEM methods, and as one would expect just the nanostructured coatings (curves 4 and 6) demonstrated the higher resistance and stability.

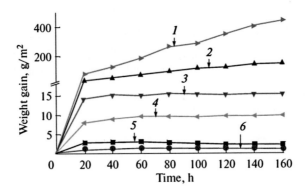

Fig. 5.10 Weight gain versus corrosion time for ASTM 1020 steel (curves *1* and *2*) and conventional coatings (those *3* and *5*) as well as nanostructured coatings (those *4* and *6*) in Na_2SO_4–30 % K_2SO_4 environment at 550 °C (curves *2*, *5* and *6*) and 650 °C (curves *1*, *3* and *4*) (adapted from [64])

In [64], it is marked out that the nanostructured coatings can be recommended for the boiler tubes protection. Other examples of some nanostructured coatings effective applications for the corrosion protection with specific aims are presented in the following publications: the Ni-based superalloy nanocrystalline protective coating in NaCl acidic solution [65]; nanostructured NiCoCrAlY coatings for oxidation protection [66]; Ni-based superalloy K36 (with Y addition) for molten sulfate at 900 °C [67]; HfC/SiC multi-layer coatings for the protection of carbon composites [68]; nanostructured CoNiCrAlYSi coatings for cyclic hot corrosion in molten salt at 880 °C up to 640 h [69]. In [65–67, 69], it was marked an accelerated formation of the passivation films-based Cr_2O_3 and Al_2O_3.

For comparison purposes in Table 5.8, the data, concerning some nitride nanoparticles (NbN, VN, CrN, and TiN) and their behavior in the various liquid media, are summarized and systematized (the error of the v value definitions is about ±1 Å/day) [70].

It is worth noting that for the VN particles interaction with HNO_3 a total dissolution is observed, but at the same time, the TiN particles reaction form oxide. The most aggressive liquid medium turns to be alkaline solution. Under the studied conditions, the highest corrosion resistance has been revealed for the chromium oxide samples, and the authors [70] are even recommending this compound for a potential usage in the proton exchange membranes of the fuel elements.

The last achievements in the field of the nanostructured titanium alloys (as well as the multi-component bioactive nanocoatings on their basis), especially for the medical applications, are well known and were described and analyzed in details in many recent publications (e.g., [71–74]). Many of these materials have already demonstrated very perspective combination of the chemical, physical, tribological, and biologic properties, and therefore, many of these coatings are used or recommended for the various medical items fabrication (such as the implants, fixers, and artificial limbs; see also early Fig. 4.19). It was found also that some calcium-phosphates coatings, deposited on the implant surfaces, are raising their corrosion resistance in the physiological solutions and aggressive media [72, 74]. The recent data have also shown that the Ti-based metallic nanoglasses (of $Ti_{34}Zr_{14}Cu_{22}Pd_{30}$ type) offer a high compatibility with the biological media [75].

Accumulation of information about the corrosion properties of nanomaterials is very intensive and some of the most interesting recent issues are listed below:

1. Stress corrosion cracking of UFG and conventional Al–7.5Mg alloy [76];
2. Oxidation of ZrB_2 composites-based various transition metal silicides (such as $ZrSi_2$, $MoSi_2$, $TaSi_2$, and WSi_2) at $T = 1200–1800$ °C [77];
3. The effect of bulk and surface SPD on corrosion behavior corrosion fatigue of AA5083 aluminium alloy [78];
4. Comparative corrosion behavior of Zr–20 % Cr and Zr–20 % Ti alloy nanocrystalline films [79];
5. Factors controlling corrosion behavior of amorphous and nanocrystalline alloys [80].

Table 5.8 Change of the nanoparticles diameter (d), estimated by XRD, and corrosion rates (v) after treatment at 80 °C during 2 weeks in H_2O and 2M NaOH, HCl, HNO_3, and H_2SO_4 solutions (adapted from [70])

Nit-ride	d_o (nm)	H_2O		NaOH		HCl		HNO_3		H_2SO_4	
		d (nm)	v (Å/day)	d (nm)	v (Å/day)	d (nm)	v (Å/day)	d (nm)	v (Å/day)	d (nm)	v (Å/day)
NbN	23	22	0.3	20	1.0	22	0.3	17	2.1	21	0.7
VN	35	27	2.8	20	5.3	35	0	diss	–	20	5.3
CrN	31	32	(−0.3)	29	0.7	31	0	30	0.3	29	0.7
TiN	28	23	1.7	22	2.1	26	0.7	oxide	–	22	2.1

References

1. Koch CC, Ovid'ko IA, Seal S et al (2007) Structural Nanocrystalline Materials: Fundamentals and Applications. Cambridge University Press, Cambridge
2. Saji VS, Thomas J (2007) Nanomaterials for corrosion control. Curr Sci 92:51–55
3. Raiston KD, Birbilis N (2010) Effect of grain size on corrosion: a review. Corrosion 66:0750005 (1–13)
4. Liu L, LI Y, Wang F (2010) Electrochemical corrosion behavior of nanocrystalline materials —a review. J Mater Sci Technol 26:1–14
5. Maurice V, Marcus Ph (2012) Passive films at the nanoscale. Electrochim Acta 84:129–138
6. Andrievski RA (2013) The role of nanoscale effects in the interaction between nanostructured materials and environments. Prot Metals Phys Chem Surf 49:528–540
7. Mahesh BV, Singh Raman RK (2014) Role of nanostructure in electrochemical corrosion and high temperature oxidation: a review. Metall Mater Trans A 45:5799–5822
8. Erb U (2010) Size effects in electroformed nanomaterials. Key Eng Mater 444:163–188
9. Gupta RK, Singh Raman RK, Koch CC (2010) Fabrication and oxidation resistance of nanocrystalline Fe10Cr alloy. J Mater Sci 45:4884–4888
10. Rashidi AM (2011) Isothermal oxidation kinetics of nanocrystalline and coarse grained nickel: experimental results and theoretical approaches. Surf Coat Technol 205:4117–4123
11. Gupta RK, Birbilis N, Zhang J (2012) Oxidation resistance of nanocrystalline alloys. In: Shih H (ed) Corrosion Resistance. InTech, Croatia, pp 213–238 (Chapter 10)
12. Manesh BV, Singh Raman RK, Koch CC (2012) Bimodal grain sixe distribution: an effective approach for improving the mechanical properties of Fe–Cr–Ni alloy. J Mater Sci 47:7735–7743
13. Rotagha R, Langer R, El-Sherik AM et al (1991) The corrosion behavior of nanocrystalline nickel. Scr Metall Mater 25:2867–2872
14. Mishra R, Balasubramaniam R (2004) Effect of nanocrystalline grain size on the electrochemical and corrosion behavior of nickel. Corr Sci 46:3019–3029
15. Kim SH, Aust KT, Erb U et al (2003) A comparison of the corrosion behavior of polycrystalline and nanocrystalline cobalt. Scr Mater 48:1379–1384
16. Luo W, Shi P, Wang Ch et al (2012) Electrochemical corrosion behavior of bulk nanocrystalline copper in nitric acid solution. J Electrochem Soc 159:C80–C85
17. Erb U (2011) Corrosion behavior of electrodeposited nanocrystal. In: Winston R (ed) Uhlig's corrosion handbook, 3rd edn. John Wiley & Sons, West Sussex, pp 517–528 (Chapter 37)
18. Zhao Y, Cheng IC, Kassner ME et al (2014) The effect of nanotwins on the corrosion behavior of copper. Acta Mater 67:181–188
19. Syugaev AV, Lomeva SF, Reshetnikov SM (2010) Electrochemical properties of nanocrystalline α-Fe + Fe$_3$C composites in acid mediums. Prot Metals Phys Chem Surf 46(1):82–88
20. Afshari V, Dehghanian C (2010) The effect of pure iron in a nanocrystalline grain size on the corrosion inhibitor behavior of sodium benzoate in near-neutral aqueous solution. Mater Chem Phys 124:466–471
21. Umoren SA, Li Y, Wang FH (2011) Influence of iron microstructure on the performance of polyacrylic acid as corrosion inhibitor in sulfuric acid solution. Corr Sci 53:1778–1785
22. Wang SG, Sun M, Cheng PC et al (2011) The electrochemical corrosion of bulk nanocrystalline ingot iron in HCl solution with different concentration. Mater Chem Phys 127:459–464
23. Gupta RK, Singh Raman RK, Koch CC (2012) Electrochemical characteristics of nano-and microcrystalline Fe–Cr alloys. J Mater Sci 47:6118–6124
24. Hoseini M, Shahryari A, Omanovic S et al (2009) Comparative effect of grain size and texture on the corrosion behavior of commercially pure titanium processed by equal channel angular pressing (2009) Corr Sci 51:3064–3067

25. Nie M, Wang ChT, Qu M et al (2014) The corrosion behavior of commercial purity titanium processed by high-pressure torsion. J Mater Sci 49:2824–2831
26. Amirhanova NA, Valiev RZ, Chernyaeva EYu et al (2010) Corrosion behavior of titanium materials with an ultrafine-grained structure. Russ Metall (Metally) 2010(5):456–460
27. Chuvil'deev VN, Kopylov VI, Bakhmet'ev et al (2012) Effect of the simultaneous enhancement in strength and corrosion resistance of microcrystalline titanium alloys. Doklady Phys 57:10–13
28. Bozhko PV, Korshunov AV, Il'in AP et al (2012) Reactive capacity of submicrocrystalline titanium. II. Electrochemical properties and corrosion resistance in sulphuric acid solutions. Perspective Mater 5:13–20 (in Russian)
29. Korshunov AV, Il'in AP, Lotkov AI et al (2012) Reactive capacity of submicrocrystalline titanium. I. Regularities of oxidation during air heating. Perspective Mater 4:5–12 (in Russian)
30. Mathur Sh, Jain R, Kumar P et al (2012) Effect of nanocrystalline phase on the electrochemical behavior of the alloy $Ti_{60}Ni_{40}$. J All Comp 538:160–163
31. Nikulin SA, Rogachev SO, Rozhnov AB et al (2012) Resistance of alloy Zr–2.5 % Nb with ultrafine-grain structure to stress corrosion cracking. Met Sci Heat Treatm 54:407–418
32. Argade GR, Kumar N, Mishra RS (2013) Stress corrosion cracking susceptibility of ultrafine grained Al–Mg–Sc alloy. Mater Sci Eng, A 565:80–89
33. Brunner JG, Birbilis N, Ralston KD et al (2012) Impact of ultrafine-grained microstructure on the corrosion of aluminium alloy AA2024. Corr Sci 57:209–214
34. Vrátná J, Hadzima B, Bukovina M et al (2013) Room temperature corrosion properties of AZ31 magnesium alloy processed by extrusion and equal channel angular pressing. J Mater Sci 48:4510–4516
35. Youssef KhMS, Koch CC, Fedkiw PS (2004) Improved corrosion behavior of nanocrystalline zinc produced by pulse-current electrodeposition. Corr Sci 46:51–64
36. Li MCh, Jiang LL, Zhang WQ et al (2007) Electrochemical corrosion behavior of nanocrystalline zinc coatings in 3.5 % NaCl solutions. J Sol State Electroch 11:1319–1325
37. Voitovich VB, Lavrenko VA, Adejev VM (1994) High-temperature oxidation of titanium boride of different purity. Oxid Met 42:145–161
38. Andrievski RA, Lavrenko VA, Desmaison J et al (2000) High-temperature oxidation of AlN-base films. Doklady Phys Chem 373:99–101
39. Lavrenko VA, Panasyuk AD, Desmaison-Brut M et al (2005) Kinetics and mechanism of electrochemical corrosion of titanium-based ceramics in 3 % NaCl solution. J Eur Cer Soc 25:1813–1818
40. Dranenko AC, Lavrenko VA, Talash VN (2010) Corrosion stability of nanostructured TiB_2 films in 3 % NaCl solution. Powder Metall Met Cer 49(3–4):74–178
41. Dranenko AC, Lavrenko VA, Talash VN (2013) Corrosion resistance of TiN films in 3 % NaCl solution. Powder Metall Met Cer 52(3–4): 223–227
42. Pan X, Shen K, Xu J et al (2012) Preparation and corrosion resistance of TiB_2 amorhous-crystalline films. Chin J Electr Dev 35:135–138
43. Barkovskaya MM, Uglov VV, Khodasevich VV (2011) Composition and corrosion resistance of coatings on the basis of nitrides of titanium and chromium. J Surf Invest 5:402–409
44. Her Sh-Ch, Wu Ch-L (2012) Corrosion resistance of TiN coating on 304 steel. Appl Mech Mater 121–126:3779–3783
45. Kuptsov KA, Kiryukhantsev-Korneev PhV, Sheveiko AN et al (2013) Comparative study of electrochemical and impact wear behavior of TiCN, TiSiCN, TiCrSiCN, and TiAlSiCN coatings. Surf Coat Techn 216:273–281
46. Bondarev AV, Kiryukhantsev-Korneev PhV, Sheveiko AN et al (2015) Structure, tribological and electrochemical properties of low friction TiAlSiCN/MoSeC coatings. Appl Surf Sci 327:253–261
47. Conde A, Navas C, Cristobal AB et al (2006) Characterization of corrosion and wear behaviour of nanoscaled e-beam PVD CrN coatings. Surf Coat Techn 201:2690–2695
48. Larijani MM, Elmi M, Yari M et al (2009) Nitrogen effect on corrosion resistance of ion beam sputtered nanocrystalline zirconium nitride films. Surf Coat Techn 203:2591–2594

49. Escobar C, Villareall M, Caicedo JC et al (2013) Diagnostic of corrosion-erosion evolution for [Hf-Nitrides/V-Nitrides]n structures. Thin Sol Films 545:194–199
50. Escobar C, Caicedo JC, Aperator W et al (2014) Corrosion resistant surface for vanadium nitride and hafnium nitride layers as a function of grain size. J Phys Chem Sol 75:23–30
51. Grigoriev ON, Galanov BA, Lavrenko VA et al (2010) Oxidation of ZrB_2–SiC–$ZrSi_2$ ceramics in oxygen. J Eur Cer Soc 30:2397–2405
52. Fahrenholtz WG, Hilmas GE (2012) Oxidation of ultra-high temperature transition metal diboride ceramics. Int Mater Rev 57:61–72
53. Carney C, Paul A, Venugopal S et al (2014) Qualitative analysis of hafnium diboride based ultra high temperature ceramics under oxyacetylene torch testing at temperatures above 2100 °C. J Eur Cer Soc 34:1045–1051
54. Gherrab M, Garnier V, Gavarini S et al (2013) Oxidation behavior of nano-scaled and micro-scaled TiC powders. Int J Refr Met Hard Mater 41:590–596
55. Kurlov AS, Gusev AI (2013) Oxidation of tungsten carbide powders in air. Int J Refr Met Hard Mater 41:300–307
56. Zhao G, Zhang X, Shen Zh et al (2014) Oxidation of ZrB_2 nanoparticles at high temperature under low oxygen pressure. J Am Ceram Soc 97:2360–2363
57. Musil J (2012) Hard nanocomposites coatings: thermal stability, oxidation resistance and toughness. Surf Coat Techn 207:50–65
58. Zhang XY, Shi MH, Li C et al (2007) The influence of grain size on the corrosion resistance of nanocrystalline zirconium metal. Mater Sci Eng, A 448:259–263
59. Ralston KD, Birbilis N, Davies CHJ (2010) Revealing the relationship between grain size and corrosion rate of metals. Scr Mater 63:1201–1204
60. Volonakis G, Tsetseris L, Logothetidis S (2011) Electronic and structural properties of TiB_2: bulk, surface, and nanoscale effects. Mater Sci Eng B 176:484–489
61. Bouzoubaa A, Diawara B, Maurice V et al (2009) Ab initio modeling of localized corrosion: study of the role of surface steps in the interaction of chlorides with passivated nickel surfaces. Corr Sci 51:2174–2182
62. Cross SR, Woolham R, Shademan S et al (2013) Computational design and optimization of multilayered and functionally graded corrosion coatings. Corr Sci 77:297–307
63. Gunasegaram DR, Venkatraman MS, Cole IS (2014) Towards multiscale modeling of localized corrosion. Int Mater Rev 59:84–114
64. Tao K, Zhou X, Cui H et al (2008) Preparation and properties of a nanostructured NiCrC alloy coating for boiler tubes protection. Mater Trans 49:2159–2162
65. Liu L, Li Y, Wang F (2008) Influence of grain size on the corrosion behavior of a Ni-based superalloy nanocrystalline coating in NaCl acidic solution. Electrochim Acta 53:2453–2462
66. Mercier D, Gauntt BD, Brochu M (2011) Thermal stability and oxidation behavior of nanostructured NiCoCrAlY coatings. Surf Coat Technol 205:4162–4168
67. Yu P, Wang W, Wang F et al (2011) High-temperature corrosion behavior of sputtered K38 nanocrystalline coatings with and without yttrium addition in molten sulfate at 900 °C. Surf Coat Technol 206:68–74
68. Verdon C, Szwedek O, Jacques S et al (2013) Hafnium and silicon carbide multilayer coatings for the protection of carbon composites. Surf Coat Technol 230:124–129
69. Yaghtin AH, Javadpour S, Shariat MH (2014) Hot corrosion of nanostructured CoNiCrAlYSi coatings deposited by high velocity oxy fuel process. J All Comp 584:303–307
70. Yang M, Allen AJ, Nguyen MT et al (2013) Corrosion behavior of mesoporous transition metal nitrides. J Sol St Chem 205:49–56
71. Kolobov YuR (2010) Nanotechnologies for the formation of medical implants based on titanium alloys with bioactive coatings. Nanotechnol Russ 4:758–775
72. Shtansky DV, Levashov EA (2013) Recent progress in the field of multicomponent bioactive nanostructured films. RCS Adv 3:11107–11115
73. Mishnaevsky L Jr, Levashov E, Valiev RZ et al (2014) Nanostructured titanium-based materials for medical implants: modeling and development. Mater Sci Eng R 81:1–19

74. Lyakhov NZ (ed) (2014) Biocomposites on base of calcium-phosphate coatings, nanostructural and ultra-fined grained bioinert metals, their biocompatibility and biodegrdation. Publ House Tomsk St Univ, Tomsk (in Russian)
75. Gleiter H, Schimmel Th, Han H (2014) Nanostructured solids – from nano-glasses to quantum transistors. Nano Today 9:17–66
76. Sharma MM, Tomedi JD, Weigley TJ (2014) Slow strain rate testing and stress corrosion cracking of ultra-fine grained and conventional Al–Mg. Mater Sci Eng A 619:35–45
77. Silverstroni L, Meriggi G, Sciti D (2014) Oxidation behavior of ZrB_2 composites doped with various transition metal silicides. Corr Sci 83:281–291
78. Abdulstaar M, Mhaede M, Wollmann M et al (2014) Investigating the effect of bulk and surface severe plastic deformation on the fatigue, corrosion behavior and corrosion fatigue of AA5083. Surf Coat Technol 254:244–251
79. Ali F, Mehmood M, Gasim AM et al (2014) Comparative study of the structure and corrosion behavior of Zr–20 %Cr and Zr–20 %Ti alloy films deposited by multi-arc ion plating technique. Thin Sol Films 564:277–281
80. Królikowski A (2015) Corrosion behavior of amorphous and nanocrystalline alloys. Solid State Phenom 227:11–14

Chapter 6
Conclusions

There can be no doubt that NMs with their unique properties will remain a major focus of interest for the material science investigations and effective innovation activities in the twenty-first century. This interest is conditioned not only by such materials with their very high mechanical, physical, chemical, and operational parameters, but also by these parameters stability under performance conditions, and the authors tried to consider and analyze this problem in their monograph. The data presented give many evidences of NMs stability under the extreme conditions including high temperatures and pressures, as well as irradiation and mechanical/corrosion actions. A high general potential of using NMs under these conditions due to the presence of numerous low-energy interfaces, nanotwinned and gradient surface structures, etc. is considered. Great perspectives of these structures and, above all, gradient objects are also highlighted in many recent publications (e.g., [1–5]). It should also be noted the continuous increasing accumulation of general nano-information that the authors sought to consider adding references to the latest publications at the end of each of the Chaps. 2–5.

To our regret, many nano-objects and materials (such as nanosemiconductors, quantum dots, nanoclusters, nanoparticles and nanotubes, fullerites, graphenes, polymers, hybrid and amorphous nanocomposites, etc.) remain mainly beyond consideration or are analyzed very fluently.

Discussing the stability problems, the authors tended to mark the poorly understood questions, and it seems useful to itemize them in short:

1. The experimental studies of NMs behavior under extreme conditions still remain short as well as insufficient, and therefore, they must be prolonged, widened, and deepened.
2. The theoretical approaches and modeling of the NMs stability mechanisms under extreme conditions must be developed, especially for the synergy effects during possible combined actions.
3. At last, it is necessary to optimize the technological processing regimes of some universal NMs capable to withstand various extreme loadings.

Alas, today the nanostructured approach application to heat-resistant materials, nuclear ones, and other advanced objects remains obviously insufficient. The

© Springer International Publishing Switzerland 2016
R.A. Andrievski and A.V. Khatchoyan, *Nanomaterials in Extreme Environments*,
Springer Series in Materials Science 230, DOI 10.1007/978-3-319-25331-2_6

authors hope that their modest book will be useful and provide some additional insights into NMs opportunities.

In his well-known lecture (1959) titled "There is a plenty room at the bottom", Richard Feynman invited scientists to a new physical world. Now, this "plenty room at the bottom" is gradually filling in by new mechanical objects, materials, and systems, creating new opportunities, challenges, and problems. The deep understanding of new fundamental features of the nanoworld allows us to create fantastic new materials for new giant projects. Much work to the realization of these plans lies ahead!

Acknowledgments   The authors are very grateful to the Professor Dr. Herbert Gleiter for kind and inspiring attention as well as to the many colleagues and friends for sending their interesting papers. Supports from the Russian Basic Research Foundation and the Russian Academy of Sciences are also appreciated.

References

1. Zheng Sh, Beyerlein IJ, Carpenter JS et al (2013) High-strength and thermally stable bulk nanolayered composites due to twin-induced interfaces. Nature Commun 4:1696 (1–8)
2. Lu K (2014) Making strong nanomaterials ductile with gradients. Science 345:1455–1456
3. Wu XL, Jang P, Chen L et al (2014) Synergetic strengthening by gradient structure. Mater Res Lett 2:185–191
4. Wu XL, Jang P, Chen L et al (2014) Extraordinary strain hardening by gradient structure. Proc NAS 111:7197–7201
5. Andrievski RA (2015) The role of interfaces in nanomaterials behavior at extremes. Diffusion Found 5:147–170

Index

A

Abnormal grain growth, 7, 10, 15, 17
Accumulative roll bonding (ARB), 55, 56, 64
Alloys:
    Al–Mg, 71
    Al–Mg–Si, 74
    Al–Zn, 63
    Cu–0.5Al$_2$O$_3$, 30
    Cu–Cr, 74
    Cu–Nb, 12, 59, 68
    Cu–Si, 61
    Cu–Ta, 14
    Cu–W, 14
    Cu–Zr, 12, 64
    Fe–C, 63
    Fe–Cr, 80
    Fe–Fe$_3$C, 84
    Fe–Zr, 12
    Ni–Cr–C, 96
    Ni–W, 12
    Pd–Zr, 15
    Sm–Co, 12
    Ti$_{60}$Ni$_{40}$, 86
    Ti–V–Al, 85
    W–Ti, 14
    W–0.5TiC, 30
    Zn–22 % Al, 62
    Zr–2.5Nb, 86
Amorphization, 27, 32, 34, 39, 64
Agues solutions:
    HCl, 85
    HNO$_3$, 83
    H$_2$SO$_4$, 82–85, 89
    NaCl, 83
    Na$_2$SO$_4$, 83, 84

B

BET, 90
Body-central cubic (BCC), 13, 41, 42, 71
Borides:
    B$_4$C, 48
    TiB, 89
    TiB$_2$, 48, 66, 88, 91, 93, 95
    Ti(B,C)$_X$, 18
    Ti(B,N)$_X$, 18
    Si–B–C–N, 94
    ZrB$_2$, 90

C

Carbides:
    B$_4$C, 18
    SiC, 32, 33, 37, 38, 48, 90, 93
    TiC, 47, 91
    WC, 91
    WC–Co, 17

D

Density function theory (DFT), 12, 95
Diamond, 17, 60, 72

E

Equal channel angular pressing (ECAP), 21,
    56, 60, 85, 86

F

Face-central cubic (FCC), 13, 41, 43, 71
Fracture toughness, 38, 59

G

Gradient surface structure, 56, 103

© Springer International Publishing Switzerland 2016
R.A. Andrievski and A.V. Khatchoyan, *Nanomaterials in Extreme Environments*,
Springer Series in Materials Science 230, DOI 10.1007/978-3-319-25331-2

H
Hall-Petch, 58, 95
High-angle boundaries, 61
High-pressure torsion (HPT), 17, 21, 56,
    62–64, 66, 74

I
Interface, 3, 35, 39, 59, 68, 79
Interstitial atoms and vacancies (IAV), 27, 32,
    39, 43

K
Kinetic approach, 9, 12, 14

L
Lamellar spacing width, 15
Low-angle boundaries, 9, 61

M
Metals and some other elements:
    Al, 87
    Au, 32, 34, 35, 43
    B, 48
    Cr, 81
    Cu, 10, 13, 34, 36, 60, 61, 72, 83
    Cu/Nb, 31
    Cu/V, 31
    Cu/W, 31
    Fe, 13
    Mg, 87
    Nb, 69, 86
    Ni, 13, 30, 41, 81, 82
    Pd, 29
    Pt, 34
    Ti, 56, 60, 63, 73, 85
    W, 11, 35, 46, 48
    Zr, 34, 95
Molecular dynamics (MD), 9, 41, 43, 71

N
Nanocomposite, 3, 11, 74
Nanocrystallite, 3
Nanoglasses, 3
Nanolaminated, 31, 57, 59, 75
Nanotwinned, 15, 59, 69, 83, 103
Nitrides:
    AlN, 19, 93
    Boron nitride (BN), 18, 21, 60, 72
    CrN, 19, 97

NbN, 97
nc-(Al, Ti)N/a-Si$_3$N$_4$, 20
Si$_3$N$_4$, 20
TiAlCrSiYN/TiAlCrN, 68
TiAlSiCN, 20
TiN, 18, 66, 89
TiN–SiN, 12
VN, 97
ZrN, 18, 19

O
Oxides:
    Al$_2$O$_3$, 66, 93, 97
    Al$_2$O$_3$–ZrO$_2$, 12
    B$_2$O$_3$, 91, 93
    CeO$_2$, 66
    Cr$_2$O$_3$, 96, 97
    Fe$_2$O$_3$, 63, 85
    (Fe, Ti, Y)O$_X$, 35
    La$_2$O$_3$, 47
    NiO, 81
    SiO$_2$, 32, 93
    TiO$_2$, 43, 65, 86, 89, 91, 93
    UO$_2$, 48
    Y$_2$O$_3$, 9, 47
    ZnO, 63
    ZrO$_2$, 29, 67

P
Primary knocked-out atoms (PKA), 27, 41, 42

S
Severe plastic deformation (SPD), 8, 55, 68
Stability map, 11
Stacking-fault tetrahedra (SFT), 17, 34
Steels:
    ASTM 1020, 96
    F95, 44, 75
    Fe–9/14/18Cr, 44
    M93, 44
    MA957, 37
    ODS 9–12Cr, 37
    ODS-EUROFER, 38
    SUS316L + 1 %TiC, 37
    14YWT, 30, 35, 37, 38, 44
Stress corrosion cracking, 86, 97
Superhard coatings (films, materials), 18, 20,
    59

Printed in the United States
By Bookmasters